The Science of

HEROES

The Science of

HER⊕ES

The Real-Life Possibilities
Behind the Hit TV Show

YVONNE CARTS-POWELL

BERKLEY BOULEVARD BOOKS, NEW YORK

THE BERKLEY PUBLISHING GROUP
Published by the Penguin Group
Penguin Group (USA) Inc.
375 Hudson Street, New York, New York 10014, USA
Penguin Group (Canada), 90 Eglinton Avenue East, Suite 700, Toronto, Ontario M4P 2Y3, Canada
(a division of Pearson Penguin Canada Inc.)
Penguin Books Ltd., 80 Strand, London WC2R 0RL, England
Penguin Group Ireland, 25 St. Stephen's Green, Dublin 2, Ireland
(a division of Penguin Books Ltd.)
Penguin Group (Australia), 250 Camberwell Road, Camberwell, Victoria 3124, Australia
(a division of Pearson Australia Group Pty. Ltd.)
Penguin Books India Pvt. Ltd., 11 Community Centre, Panchsheel Park, New Delhi—110 017, India
Penguin Group (NZ), 67 Apollo Drive, Rosedale, North Shore 0632, New Zealand
(a division of Pearson New Zealand Ltd.)
Penguin Books (South Africa) (Pty.) Ltd., 24 Sturdee Avenue, Rosebank, Johannesburg 2196, South
Africa

Penguin Books Ltd., Registered Offices: 80 Strand, London WC2R 0RL, England

While the author has made every effort to provide accurate telephone numbers and Internet addresses
at the time of publication, neither the publisher nor the author assumes any responsibility for errors,
or for changes that occur after publication. Further, publisher does not have any control over and does
not assume any responsibility for author or third-party websites or their content.

PRINTING HISTORY
Berkley Boulevard trade paperback edition / October 2008

Library of Congress Cataloging-in-Publication Data

Carts-Powell, Yvonne.
  The science of heroes : the real-life possibilities behind the hit TV show / Yvonne Carts-Powell.
    p. cm.
  Includes bibliographical references.
  ISBN 978-0-425-22335-2
  1. Science—Miscellanea. 2. Human physiology. 3. Parapsychology. 4. Heroes (Television
program) I. Title.

  Q173.C317 2008
  500—dc22

                                                                    2008019836

PRINTED IN THE UNITED STATES OF AMERICA

10 9 8 7 6 5 4 3 2 1

For Sylvia Carts,
my mother,
who was also my first science teacher and storyteller.

Acknowledgments

A lot of people helped me with this book. In particular, I very much appreciate the fascinating information supplied by Dr. Amy Chused, Phill Nimiskern, Jeff Bigler, Professor Sergio Fantini, Dr. Peter Bergethon, Michell Charity, Regis Donovan, Vicka Corey, Lenny Foner, and Elizabeth Ryan.

Thanks also to a varied group of early readers, including Megan Morris, Sarah Booth, Dr. Amy Chused, bell, Cynthia Shettle (whose knowledge of *Heroes* dwarfed my own), Kyla Mackay Smith, Andy Bressen, Laura Waterstripe, Rich Macchi, and Elizabeth Liolin, and on Livejournal: Hth_the_first, Purna, and Thefourthvine. In particular, thanks to Georgina Paterson, for help above and beyond the call of friendship.

This book would have been far more difficult to write without the aid of the public libraries of the towns of Belmont and Arlington, Massachusetts, and the ever-fascinating selections from the Harvard Bookstore.

Great thanks are also due to Rickland and Rex Powell, Steve Pasechnick, my parents, and Valerie King for putting me up, and for putting up with me, while I wrote this book.

Contents

Chapter 1

THIS IS HOW SCIENCE WORKS

Ordinary people with extraordinary powers are at the center of NBC's hit show *Heroes*. They are people gifted with and endangered by forces they don't understand. People just like us.

All of us, singly and as the members of a civilization, grapple with forces we don't understand. Sometimes, it's merely a matter of understanding a human-built system: how do we fix a flat tire, or register for classes at college, or deal with a difficult boss? Other times, it's a matter of understanding the forces of nature: why a tire goes flat, why houseplants thrive (or don't), why we catch colds, why the world as a whole is getting warmer—in short, how reality works. The universe is huge and complex. We have discovered many of its underlying rules, and sometimes the things we discover about it are just amazingly strange.

On *Heroes,* our favorite characters try to understand their parents and come to grips with the shadowy Company. They are trying to escape, or find, or capture the murderous Sylar. They also try to control their superpowers and incorporate seemingly fantastic abilities into their view of what is possible in the world.

"Truth is stranger than fiction; fiction has to make sense." So said humorist Leo Rosten, and he wasn't the first to have that idea. Reality just has to be demonstrably real. We are left to try to make sense of it. Like our characters, there are still plenty of things that we don't understand at all: What is time? How do the physical rules of huge things and tiny things connect? What is consciousness? How do we think? Why and when did we begin to think? How did life begin? Is there intelligent life elsewhere in the universe? What is death? Why is death?

Still, that's the human condition: we are forever fumbling forward in the dark, trying to spread some light, discover meaning, make some progress toward our goals. That's not what this book is about—or at least, this book is about only a subsection of the grand human endeavor. It is about the questions that can be answered with science.

We have a tremendously powerful tool for investigating our universe. That tool is the scientific method. Scientific inquiry is not the only way of learning, but it is unmatched in the rigor of its questioning and the success of its conclusions.

The conclusions formed using the scientific method form a solid, dependable basis for the technologies that make our daily lives possible, including the electronics that connect us via the Internet and cell phones, the electrical grid and gas distribution systems that keep our lights on and our cars running, the medical tests and devices that saves lives. A middle-class American today has far better quality food, clothing, heating, and plumbing than even the Sun King, France's Louis XIV, one of history's greatest and most extravagant rulers, who lived three centuries ago.

SCIENCE IN A LAPTOP

Think you know all the ways technology affects your life? Bet you don't. Take a laptop computer, for example. The technological developments that make a laptop computer possible are many, and varied, and amazing. And behind each technological engineering advance, there exists basic scientific research that gave the engineers the tools they needed before I could sit here and type, or watch a YouTube video, listen to some MP3s, send e-mail to my cousins several thousand miles away in another country, or engage in that pinnacle of modern frivolity: playing FreeCell.

Today, I sit in a coffee shop, writing on a laptop computer with a case made almost entirely of plastics developed since the 1950s, running on transistors developed in the 1970s and photolithographic circuits developed in the 1980s. The data storage mechanism (the hard drive) is based on a materials property discovered by scientists and named giant magnetoresistance in the late 1980s

. . . *continued from pg 5*

and commercialized in hard drives by the late 1990s. I read the text on a screen made from liquid crystals, which were the subject of research for almost a hundred years before the first liquid crystal displays were patented and developed in the early 1970s.

The CD that I burn to back up my data is made from polycarbonate, first documented by a material scientist around the turn of the last century, and first commercialized in 1960—the same year that the first laser was demonstrated, but decades before the development of the diode lasers that were compact and efficient and cheap enough to put into an optical drive that encodes data on the CD.

Ben Franklin was experimenting with methods of storing electricity well before Alessandro Volta made the first thing we would consider a battery over two centuries ago. Volta's was a wet-cell battery, however, whereas the battery running my laptop is a dry cell, which took another eighty years to develop. My battery is also based on lithium—research on that sort of material for batteries started a little over a century ago, and the lithium ion battery was commercialized only a decade ago. (And, given the recent reports of lithium ion batteries catching fire, it's clear that we're still in the process of developing them.)

I'm checking an encyclopedia via the Internet as I type, using my Wi-Fi connection. Wi-Fi is a wireless data communications system based on a standard developed only ten years ago by the Institute of Electrical and Electronics Engineers. As you might expect, this nonprofit organization has a lot of work to do—and it does it, with more than 370,000 members, running hundreds of publications and conferences as well as developing and proposing well over a thousand standards. (Full disclosure: I'm a member of the IEEE.)

What a piece of work is a laptop, how infinite in applications! As long as it's working correctly, the technology that made it possible is transparent to me, the user. And I don't need to know the science behind the hard drive or the screen or the keyboard or the wireless connection in order to use the laptop. But in order to continue reaping the benefits of science and technology, we need scientists and engineers and a society that values both science and engineering.

One benefit based on my laptop might be delivered by the "One Child One Laptop" association, which is trying to develop very cheap laptops for schoolchildren around the world. Their goal is to provide children with new opportunities to explore, experiment, and express themselves by developing very cheap laptops running nonproprietary software. The goal is to make $100 laptops, which could be bought in bulk by governments for their school systems. The group hasn't gotten the price down to $100 yet (they cost about $190 each, as I type this). A commercial company (Asus) has introduced laptops that cost less than $250 each. This may be bad for the association, but it's great for the students.

My laptop is a marvel. But it's a short-lived one: today's laptops have expected life spans of about five years, by which time even more new technologies will, the makers expect, have made my current machine obsolete. And five years would be positively ancient for a cell phone—their expected lifetimes are eighteen months.

Fruit of the labors

This book is about the fruit of the labors of scientists and engineers. While the dreamers and doers in the world of *Heroes* are trying to save the world one cheerleader at a time, the real world is also full of natural wonders and technological marvels that rival what we see on the screen. And we will need more science and technology, as well as politics and grassroots efforts, to save us from the global challenges we face, like the warming of the earth due to greenhouse gases, degradation and destruction of the environment, and the lack of clean energy and clean water and education for billions of people.

Making a scientific or engineering breakthrough is a little like discovering a superpower. Our abilities increase when we develop a better understanding of the universe or develop better tools for manipulating our environment. As just one of many examples, the first microscopes let people see the bacteria that caused disease, which allowed Pasteur to develop ways to make milk and other liquids safer to drink. These breakthroughs can be as exciting, even if not quite as personal, as learning how to fly.

There's another difference, though. Hiro's superpowers are (so far as we know) unique to him. It seems that super-powers are like artistic expression: one person, one power, just as no two painters would create the same painting, nor two poets the same haiku. Science is less idiosyncratic. If one scientist doesn't uncover a facet of reality, then chances are that another one will. Reality is just reality (whether we understand it or not), and if a number of people are investigating it with the same tools, then multiple people can discover something independently. That's a comforting thought. If we didn't have Beethoven, then nobody would have written the "Ode to Joy," but if we didn't have Darwin, then someone

PROVING—AND RE-PROVING—THAT THE WORLD IS ROUND

The first experiments (for which we have records) that determined the shape and diameter of the earth were done by Eratosthenes in roughly 200 B.C. His experiment can be replicated by anyone willing to dedicate a stick, a ruler, and a compass to the task. Each year, hundreds of schoolchildren repeat his research. His idea was conceived, tested, lost, rediscovered, retested—and very possibly independently conceived—over the last two thousand years.

Without the basic idea of the earth being round, Christopher Columbus would not have sailed west to find a route to the Far East. Nor would Ferdinand Magellan's ship, the *Victoria*, have even attempted to circumnavigate the world, thus experientially proving that if you headed west from Spain, you ended up in the Far East. (Magellan didn't make it all the way around the globe on that trip, though. He was killed in the Phillippines. The Warner Bros. television show *Animaniacs* has a surprisingly accurate song about his misadventures.)

Without the basic idea that the world is round, the Europeans of the Renaissance could not even have begun to argue about whether the earth rotates around the sun, or vice versa. The former was a dangerous idea to the powerful Church that had a lot invested in the idea of being the center of God's creation. But despite powerful political pressure, common consensus eventually led to a recognition of the shape of our solar system, which led to the intellectual tools necessary for space flight.

A few people today still believe that the earth is flat. They might be merely ignorant, but many of them are willfully ignoring both reasoning and evidence. Nothing—not even common sense—can force a person to believe something, no matter how self-evident it seems. People are perfectly capable of ignoring the fruits of science, disbelieving and disavowing the proof of their own senses, and throwing away the benefits conveyed by science and technology—and, for that matter, arts and culture and legal systems.

else would have looked at the same evidence that he did and developed essentially the same theory of evolution.

The first person to create and use calculus held in his hand a powerful tool for describing reality—a tool so powerful and important that it has become a regular course in our schools. Isaac Newton usually gets the credit for inventing calculus, and he certainly put the method to immediate use, but his contemporary Gottfried Wilhelm Leibniz not only developed the method of calculus independently but also came up with the symbols we use for it today.

Our civilization is based not just on laws, and culture, and our frankly magnificent technology, but on our ideas of how the universe works and our place in the universe. And while that job starts with philosophers, the place where the rubber hits the road—where ideas are tested and thrown out if they don't work or developed if they do—is in the practice of science.

There are rules to testing theories. It's not enough to say, "No, I don't like that idea, so I don't accept it." It's also not enough to say, "Someone I trust says that isn't true." A scientist is a little like a judge in that there is no objective truth: there is some evidence, and competing arguments that must be winnowed through in order to find the most likely explanation. The argument doesn't have to agree with other theories about the world, but it does have to be consistent with the observations that we have. Once an explanation seems viable, it's time to see if it can accurately predict the outcome of experiments. And then the fun starts: handing it over to the rest of the world for further testing.

Just like compound interest in a bank account, scientific knowledge accumulates. Isaac Newton wrote to a friend, "If I have seen further [than certain other men] it is by standing upon the shoulders of giants." He was referring to the work of Kepler and Galileo. I am no smarter than my many-times-great-grandmother who lived at the same time as Newton, but I have access to much much more knowledge, a much deeper view of the universe's workings, than she did, merely because I get to stand on the shoulders of more giants. We all do.

Seeing farther is a fine thing. What we do with what we see is another matter. The engineers, the people who understand and apply those rules to create new technology, have created many of the marvels and some of the terrors of our civilization. We have fantastic new medicines today. We also have poison gas in a Tokyo subway. Nuclear medicine has saved my uncle's life by destroying his cancer. Nuclear weapons threaten us all. Once some new aspect of reality is discovered, its application is a profoundly ethical affair.

Despite the danger, the search for new knowledge continues. People poke and prod and theorize all the time. Although Claire's and Matt's new powers are extraordinary, the way they test their new abilities is very ordinary. Most of us would do the same. Curiosity may have killed the cat, but cats have nothing on humans when it comes to investigating our universe. It's what we do. By and large, though, the fruits of science and technology have brought us advantages and progress.

While the path of scientists and engineers includes coming to some understanding of their capabilities and the world around them, self-realization is seldom as simple as saving a cheerleader, or defeating a horrible serial killer. We can't teleport like Hiro, but teleportation experiments in labs have been successful. We can't disappear like Claude, but a squid-like creature called a cuttlefish can. And we can't fly like Nathan, but we can fly in airplanes just by buying a ticket and getting through airport security. (As I write this, I've left my comfortable coffee shop to travel across the continent. Now, I'm cruising at thirty thousand feet.) People have even made human-powered planes.

If you watch television or use indoor plumbing, you are already enjoying the fruits of science and technology. But it's probably more fantastic than you imagine. Many of the superpowers in *Heroes* are not that far beyond our reach today.

The art of the story

When I say that, I don't mean that people like Hiro or Claire or Peter or Sylar live among us today. There's an excellent reason why superhero stories (and before that, fables of gods and supernatural creatures) are so popular, which has nothing to do with how realistic they are. By using fantasy, storytellers can illuminate the human condition. By telling stories that are blatantly untrue in their particulars, we can better describe deeper truths.

Why do I bother even mentioning this? Because fiction is a useful and powerful tool for artists. But while science requires imagination (after all, those ideas have to come from somewhere), it also requires testing, and one other thing. It's crucially important that the universe be consistent.

On *Heroes,* we expect a certain amount of consistency. If Sylar acts like a lying, conniving, murderous villain in one episode, we expect him to be the same sort of person in the next episode. If Hiro turns from a bouncy cubicle-dwelling fan boy into a badass warrior, we expect to see and understand both the reasons for the progression and the heroic journey that he follows. A deus ex machina ending, in which a previously unknown hero with a previously unknown gadget swoops in to save the day, isn't very satisfying. Of course, we don't know everything that's going on in the *Heroes* universe, but the show's makers should show us enough to give us a satisfying story.

The real world is a lot like that, too. We don't know everything about the universe, or our planet, or the life on Earth, or our human bodies, and we're barely scratching the surface of our human minds. We do, however, expect things that we've observed to be consistent to *stay* consistent: the sun rises in the east; it takes a year for the earth to travel around the sun; people are born, grow up, love and live and interact, and eventually die. A clear sunny sky is blue. Break a window and it will remain broken. Drop a glass of milk and it spills. Gravity works every day—if it doesn't, then Einstein was wrong and we have a very flawed idea of what gravity is. We believe, based on our human observations, that the universe is orderly. The universe obeys laws. We don't know all the laws, and indeed some of the rules like chaotic behavior and quantum uncertainty and the concept of infinity are difficult for human brains to comprehend.

Nonoverlapping magisteria

The universe contains myriad wonders, but is also self-consistent. I'm not interested in miracles. Miracles do science no good, because a miracle is not reproducible.

Miracles may exist. But science is not an appropriate tool for studying or discussing them. In his book *Rocks of Ages,* the late Harvard University professor Stephen Jay Gould developed the idea that science and religion are nonoverlapping magisteria, two areas with distinct tools for useful discussion. Religion is, he says, the correct way to ask and answer questions of ultimate meaning and moral value. Science does not—and science cannot—trespass on this area. Nor can science define art. Science doesn't tell us what makes a good story, a delightful painting, or an excellent television show (and Neilsen ratings be damned, because popularity is different from excellence).

What science does, and does very well, is answer questions about the physical world: What is stuff made of? How do larger or smaller bits of the universe work? Compared to the mysteries of the human spirit, those questions are pedestrian, and the methods used to determine the answers require slow slogging and limited certainty in the results. Nevertheless, one of the grandest of human endeavors involves watching some aspect of the universe, making a guess about a natural law based on those observations, and testing that guess.

So how does science work? It's fairly simple in concept, and it's something that humans mostly do automatically—but in practice, it is usually messy and complicated and hedged around with examples of human failings as well as illuminated by magnificent feats of human logic. Scientific experimentation isn't some abstract process: babies do it every day, as they discover first that toes exist, then that their toes are always around, then that their toes are connected to them, and finally that they can control their toes.

A baby is tremendously busy, making a mental model of herself and her world. Although the problem is more focused, this model making is much the same thing done by a troubleshooting car mechanic or a programmer. First observe, then take a guess at how something works, make a mental model that explains what you observe. We have plenty of theories—gosh, we have a lot of theories and philosophies; lack of imagination is not a human problem.

People who make their livings as scientists do the same thing, more formally. They watch, and then guess, at the behavior of groups of people, or plants, or combinations of molecules, or the motion of the stars. And they refine that guess using logic and experiments, until it is as specific and robust as possible.

Here the process sometimes bogs down, because people fall in love with their idea and may prefer to keep the idea pristine rather than test it out. But the next step is important: that guess needs to be set free to stand or fall on its own. It can be presented to other people via a master's thesis, or a presentation at a conference, or a paper published in a peer-reviewed journal.

Peer review is a process that weeds out nonsense by letting other people in the field (who are often competing with the authors for grant money and status) scrutinize the paper and agree that the paper contributes something to what we know about the subject.

Then a lot of debate goes on. Does the group's reasoning (which includes logic and almost always includes math) stand up to scrutiny? What about the experiments they did—is the design good, or is it flawed by assumptions and perhaps by limitations on funding or equipment or time? In other words, does the experiment really test the theory? And if it's a good experiment, can it be repeated by other people with substantially the same results?

This process provides self-correction, as well as a way to weed out mistakes and frauds. Take cold fusion, for example. If a fusion reaction could be created without requiring high temperatures, it could provide a source of plentiful and cheap energy—and might well win the scientists involved a Nobel Prize. In 1989, two researchers (Stanley Pons and Martin Fleischmann at the University of Utah) held a press conference reporting that they'd demonstrated cold fusion. Within days, other researchers around the world were doing their best to replicate the experiment.

Only, reproducing the experiment turned out not to be so simple. Pons and Fleischmann reported that the experiment produced more heat than could be explained by mere chemical reactions (thus indicating fusion). Occasionally, other researchers saw unexplained increases in the amount of heat generated by the equipment. But mostly they didn't. And no guesses successfully explained what triggered the increase in heat, what ended it, or whether it would occur in one situation or another. Moreover, the theories that work for other types of fusion don't predict any sort of fusion occurring in a setup like the one the original researchers suggested.

There are people still working on cold fusion today, although funding has been cut back. But until someone presents the community with a successful predictive model or a readily reproducible experiment, the existence of cold fusion will not be accepted.

The system isn't perfect. But it can produce robust ideas. The strongest theories have been examined and tested by many people. Science takes advantage of envy and competition, because if some of the people testing the theory hate the guts of the person who proposed it, the testers do their best to find legitimate problems with the idea. If they examine the theory and experimentally test the theory, and fail to either disprove it or produce a theory that explains experimental data even better, then even if we still don't accept it as the ultimate truth, it's a good enough theory to work with and build on. Then we have a guess that is not, in itself, a law of the universe, but as good an explanation as we are capable of making at this time.

That process—of watching and guessing and testing and refining—is science. And the results are durable. In math, we're still using the Pythagorean theorem, developed at least 2,500 years ago. And not only are we still telling the story about Archimedes jumping out of his bathtub and running naked down the street yelling "Eureka!" when he figured out how water displacement and weight can be used to determine an object's density—even now, after 2,300 years, we're still finding it useful, and the story (however old) is still funny.

Speaking of writing, in 1687, Newton wrote an essay about how gravity works, and the fundamental laws of motion (have you ever heard: "Things at rest tend to stay at

THE LEGEND ABOUT ARCHIMEDES AND WATER DISPLACEMENT

Long, long ago, a philosopher named Archimedes lived in Syracuse, in Sicily. He was commissioned by the king to solve a real stumper of a problem.

The king had hired a craftsman to make him a new crown, and provided a pound of high-quality gold for it. The crown was made and delivered, but the king was troubled. Had the craftsman used all of the gold? Maybe he had used less gold, and pocketed the rest? That would give the craftsman a larger profit, and human nature being what it is . . . well, the king remained unsure about whether or not he'd been bilked.

The first thing that Archimedes did was weigh the crown. It weighed the correct amount, which was a point in favor of the craftsman's honesty. But maybe the craftsman was just clever, and had mixed in some heavy but less valuable substances to the mixture, to make up the weight.

Archimedes pondered.

The king didn't want him to damage the crown. But weight alone didn't provide him with all the information he needed.

Archimedes climbed into his bath. As he sat down, the water level rose, and some splashed out. This, he realized, was the solution to his problem! He could measure the crown's volume by dunking it in a bucket and measuring the amount of water it displaced. Then, if he knew the crown's weight and its volume, he could calculate its density. If the density of the crown was the same as the density of the gold provided by the king, then the king could be pretty sure that the crown was made as specified and that the craftsman was honest!

Archimedes was so excited by this realization that he jumped out of his bath and ran to tell the king, yelling "Eureka!" (that is, "I have it!"). In his enthusiasm, he didn't bother with trivial details like putting on clothes before running out onto the streets of Syracuse.

Great story, huh? Actually, Archimedes wrote, in *On Floating Bodies,* that a body immersed in a liquid is buoyed up by a force equal to the weight of the liquid it displaces. So far as I know, he didn't write anything about kings, crowns, or episodes of streaking.

rest"? That's Newton) using calculus, in his *Principles* (or, more properly, since he was writing in Latin, *Philosophiae naturalis principia mathematica*). The rules for motion that he laid out aren't absolutely correct—but they are good enough that NASA still uses them to calculate trajectories for satellites.

By the way, you can order a translation of the *Principia* from Amazon, today. Or, given that it's long out of copyright, you can download a scanned version of the text from the Internet Archive. Personally, I wouldn't recommend going back to the original because the text is heavy going. But the ideas are still valid and still useful. Now, *that* is enduring knowledge.

Scientists aren't priests: they don't receive knowledge already cut-and-dried, from an Almighty Creator, nor should they proselytize—if anything, they should lead listeners through the same steps of reasoning and show them the same experimental results that they have gone through. But there is one assumption that underlies the entire process of science: we assume that the world is consistent across time and across space.

Today, when I kick a clod of mud, it flies through the air along a trajectory that I can calculate. (The path depends on how hard I kick it and in what direction, the force applied by gravity to pull it down, the force applied by air resistance to slow it down, etc.) We assume that when a dinosaur kicked up mud, it flew through the air in the same way that my clod of mud flies through the air today. Furthermore, we assume that if there is an alien on the other side of the universe, and if the alien kicked a clod of mud, then that mud would also obey these same rules.

So the rules that we're pretty sure are facts are sometimes called laws but more often called theories.

The guesses that Isaac Newton came up with are called "laws," today. But those laws apply only in particular situations. The guesses that Albert Einstein came up with are still called theories (like the Theory of General Relativity), but they apply to more of the universe than Newton's laws of motion. Evolution is called a theory, but it has been tested for a century and a half, and no observations or experiments thus far have been able to disprove it.

The guess is this: all life on Earth today appears to have evolved from common ancestors reaching back to a simple single-celled organism almost four billion years ago. The guess explains a whole lot of different observations. It explains the sequence and changes in the fossils that we find in rocks. It also explains why the huge variety of plant and animal life all contain chemistry that is remarkably similar. And it explains the changes we see in short-lived organisms—most notably antibiotic-resistant germs. A lot of the details about how it works are still open to refinements, but evolution itself is well established. It also provides the framework on which most of biology is based.

All our human experimenting has led us to conclude that the universe is stranger and more elaborate than we know. The universe is our playground and laboratory, exploring it and deciphering it is a grand adventure, and breakthroughs in science are grand contributions to civilization.

Science and Heroes

Both Chandra Suresh and his son Mohinder are scientists. Chandra developed a theory that genetic changes in humans were yielding individuals with varied and incredible superpowers. To test his theory, he developed an algorithm to find individuals with these abilities, and then went to New York to find them. After his father's death, Mohinder acts as both a scientist and a detective, trying to reproduce his father's work, trying to continue his work in finding superpowered people, as well as trying to solve the mystery of his father's murder.

That serves the story being told on *Heroes,* but a scientific view could go further. Mohinder's life gets complicated and dangerous and he doesn't have time to either marvel at or investigate the superpowers of the people he meets. This book looks a little more deeply at the superpowered characters on *Heroes* and at the real-world science and technology that applies to their cases.

Further reading

Eratosthene's experiment is explained at http://www.youth.net/eratosthenes/.

Stephen Jay Gould, *Ever Since Darwin: Reflections in Natural History*, Norton, 1977. Read anything by Gould—he was a marvelous writer.

Richard Feynman, "Cargo Cult Science." This was the CalTech commencement speech in 1974, about science, pseudoscience, and how to tell the difference. It is both entertaining and still very much worth reading. Available online at: http://calteches.library.caltech.edu/51/02/CargoCult.pdf.

Richard P. Feynman, Robert B. Leighton, and Matthew Sands, *The Feynman Lectures on Physics*. Three volumes. Feynman was one of the more talented and charismatic physicists of the twentieth century and a gifted teacher.

Chapter 2

ARE YOU ON THE LIST?

There wouldn't be much plot to *Heroes* if none of the characters had superpowers. And the list compiled by Mohinder's father, the late Chandra Suresh, is important for the plot. Papa Suresh, we are told, found these "evolved humans" through work with the Human Genome Project. In his book, he theorized that evolutionary changes could produce people with abilities to read thoughts, fly, heal, turn invisible, or see into the future.

In narrative terms, this is a great premise for a science-fiction story, and *Heroes* runs with the idea to produce novel situations for exploring traditional themes about heroism and responsibility, about family and power, all while introducing characters whom we love, as well as ones we love to hate. The world of *Heroes* includes families and alliances and shadowy organizations and lots of the complexity of real life. It's a fantastic story.

Scientifically, though, there are some problems with this premise. To understand why, we need to understand inheritance, what the Human Genome Project did, and how mutations and evolution work. While we're at it, let's take a look at what a virus is.

What do we know about the powers?

What are the superpowers that our characters display? What do they have in common? We see that a mysterious character, the man who is still known only as "The Haitian," can mentally block other characters, keeping them from using their superpowers. We know that the shadowy Company currently run by Bob also has drugs that can block superpowers.

We know that the brain has something to do with superpowers: Sylar is perfectly willing to kill to obtain the brains of superpowered people. But when Eden shoots herself in the head in "Fallout," Sylar cannot obtain her superpower of compulsion. We see that neither Claire nor Peter heals when they get wood or glass stuck into the backs of their skulls.

As an aside, we also know that effort—both physical and mental—requires fuel. No matter whether the effort is Claire's cheerleading acrobatics or her healing, she needs a good diet and enough calories to sustain the effort. So do the other superpowered characters: no matter whether they are erasing someone's memories (which is the Haitian's other superpower) or walking through walls or flying—that's going to take fuel. Assuming that the superpowered people on the show have the same sort of metabolism as everyone else, they probably need more calories in their diet when they use their powers. Claude, who uses his power pretty much continuously, is probably a chowhound, and we can make the same basic assumptions about Peter and Sylar, who have multiple powers.

Back to the list

We know another fact about the superpowers: in several families, multiple members are affected, which suggests that the gift of a superpower is inherited.

Before he was murdered, Chandra Suresh wrote about "activating evolution" and about "unlocking human genetic potential." He was in New York searching for people who had evolved to have superpowers like flight or telepathy. Later, Mohinder searches for the meaning of his father's life and tries to understand the conclusions of his research. He seeks out the people with superpowers whom his father was looking for, as well as trying to understand how his dad found those people. In part, he replicates his father's work as a way to understand the reason why Papa Suresh was murdered. At the root of his quest, though, Mohinder is investigating his father's work to discover who he, Mohinder, is by discovering what he has inherited from his father.

We all do this. Don't all adolescents try to define themselves by looking at how they are similar to and different from the people around them, including their families? If the highest wisdom is to "know thyself," then isn't this quest to understand ourselves through our heredity a common human part of defining ourselves?

Mohinder is better equipped to understand heredity than most: as a geneticist, he understands what we inherit from our parents on a molecular scale.

What is inheritance? Genes and DNA

Your experiences and environment shape both your personality and body, but a lot of who you are depends on the physical stuff that you inherit from your parents at your conception. You get one small cell from your dad and one rather larger cell from your mom. And that's pretty much it. When those two cells fuse at conception, all the instructions for building a human body are contained in this tiny package. They will interact with the environment, both within and beyond the womb, to influence your looks and personality, your talents and susceptibility to diseases.

Unless you have an identical twin, you can assume that your set of instructions is unique. Even identical twins, though, have different experiences starting before birth. It would be interesting to meet a pair of identical twins on *Heroes* and see whether their superpowers are the same. Alas, fraternal twins Maya and Alejandro Herrera are probably as close as we're going to get. (Because they are fraternal twins—that is, the result of two eggs being fertilized at the same time, Maya and Alejandro are no more genetically similar than any other pair of siblings, such as Nathan and Peter.)

You may have seen the motivational posters about how each person is as unique as a snowflake. That's both true, and untrue. On the one hand, each person is far more distinct than any snowflake, because each person is far more complex than a frozen water crystal could possibly be. On the other hand, our instructions, our genetic codes, are so similar—not just to other humans but to all living things— that from a chemical point of view, all the huge variety of living organisms comes down to no more than minor variations on a theme.

THE COMMON THREADS OF LIFE

One of the basic tenets of evolution is that all life descended from a common ancestor. This could be one individual creature or several life-forms of the same sort. In his *Origin of the Species*, Darwin mentioned both possibilities.

The biochemistry of all life on Earth is pretty similar. There's actually debate over what constitutes "life" with some simple structures like viruses and prions. But in general, an organism needs to replicate itself and eat and grow in order to be considered alive. Other than fulfilling those requirements, there's no need for living things to resemble one another. But they do. We can't figure out any reason why alternative forms shouldn't exist—it's just that we haven't found any (and biologists have been looking).

Every yeast, every bacterium, every animal and plant and fungus that we have encountered runs on very similar biochemical machinery. They are all based on a particular type of molecule called amino acids. There are a lot of different amino acids, but living organisms use only twenty of them. The metabolic pathways (in other words, the chemical reactions that organisms go through in order to grow) are also similar across vastly different life-forms. All forms of life use one particular chemical (adenosine diphosphate, usually called ATP) for energy.

From the simplest, smallest viruses on up through the most ridiculously complex organisms such as, for example, us, the entire range of life uses almost the same chemical process: amino acids are assembled into one particular, wonderful molecule with a distinctive 3-D form: deoxyribonucleic acid (DNA). This very long molecule in the standard double-helix shape has a backbone plus pieces bridging the gaps between the two strands. Those crosspieces are called base pairs.

A molecule much like DNA, called RNA, copies parts of the DNA and takes it over to the ribosome, where it is used to make proteins. (Okay, a few viruses skip DNA altogether, and just use RNA, but that's the only exception that I know of.) Proteins do the heavy lifting work in living organisms: they form structures, they control chemical reactions as enzymes, they carry messages between cells.

The building blocks of life are the same despite the final appearances. If you imagine a metaphorical Garden of Eden, then inside Adam and Eve and the Snake and the Apple are cells that all use the same mechanism for transporting information, processing fuel, and making copies of themselves.

To take this back to *Heroes* for a moment—do you remember the algorithm that Mohinder discovered running on his father's computer? Chandra Suresh used it to make the list of people with superpowers, and he hid a USB drive containing a copy of it in Mohinder-the-lizard's terrarium.

We never find out very much about how it works, but from Mohinder's description, it compares DNA against some formula, perhaps looking for a specific sequence of base pairs. As Mohinder says, there are more than three billion base pairs in human DNA, so there's a lot of computing that would need to go on.

The screen shots that we see appear to confirm that the algorithm is searching through base pairs, since it shows a sequence of dashes and the letters *A, T, G,* and *C.* In genetics-related research, those letters are most famous as the bases adenine, thymine, guanine, and cytosine that pair up across the center of DNA's double helix. (The mysterious symbol that appears on the screen in the pattern of dashes can only be poetic license on the storyteller's part: a different arrangement of columns would provide the same information but without the symbol.)

Not snips or snails or puppydog tails

The variations in our genes, however, determine whether your hair is kinky or smooth, whether you are a human or a dolphin, whether you are a multicelled organism with a nucleus in each cell or a simple, nucleus-less prokaryote. You are unique not just in personality but right down to your individual cells, to within each cell, to the chromosomes bundled into each cell's nucleus.

So your genes are the instructions for building you. From a biological point of view, what are you made of? Your body is made up mostly of cells or things that your cells secrete (like insulin, or digestive enzymes, to name only two of many examples). Every bit of your heart and brain and spleen, every white blood cell, every rib, and even your pinkie toe carries within it many copies of a description of you, yourself. Every one of the trillions of cells in your body carries a surprisingly full representation of you. And as you shed dead cells and build new ones, this description is copied over and over and over again.

This description determines your sex and the color of your eyes and whether you are likely to go bald. It is one factor (although not the only one) in determining how tall you are, how likely you are to get cystic fibrosis, or how good you are at doing math in your head. But (despite the news reports) there is no specific "math gene" because genes don't do math. Genes have one job, and they're good at it:

Genes are pieces of DNA that give instructions for making proteins. Your average cell has twenty-three long molecules of DNA, called chromosomes. The DNA is tightly wound up into spools. I'm told that the DNA in just one of our cells would stretch for six feet! The DNA contains instructions for making proteins, as I mentioned above, but also for how much protein to make, and for controlling the protein in other ways. There's lots and lots of feedback.

PTOOY GOES THE NUCLEUS

Your chromosomes live in the nuclei of your cells, and all of our cells have nuclei—or at least, they all start out with nuclei. We have several hundreds of specialized cells in our bodies, and a few of them don't need their nuclei to do their jobs.

Red blood cells do this. Their job is to carry oxygen from our lungs to the rest of the cells in our bodies. The more oxygen they can carry bound up to the hemoglobin protein, the better. Once the red blood cells mature, they expel their nuclei—chromosomes and all—and they become more or less sacks full of hemoglobin pushed around by the circulatory system.

In that case, a nucleus is just an inefficiency. But for some other cells, a nucleus interferes with the very purpose of the cell. For some cells in our eyeballs, transparency is a major requirement. In Mueller cells, the presence of a nucleus would scatter light unnecessarily—therefore, they operate without one.

The cells without nuclei are like worker bees: very useful but unable to replicate. They are dead ends.

I mentioned that all of life is a minor variation of the genetic code. It's a bit of an understatement to say that your genes describe you and *also* describe what species you are. In fact, *almost all of your genes describe what species you are.* Like Dr. Seuss's star-bellied sneeches, we get so caught up in noticing minor variations among humans that we fail to notice how very similar one human is to another. Each of us carries a DNA sequence that is 99.9 percent identical.

Most of our instructions dictate that we all grow livers, develop hands rather than fins, toenails instead of claws—and that's just gross physical structure. Our genes also give instructions for running the biological machinery of all life plus the specifics of the animal, like the specific instructions for keeping the soup of our innards balanced in the precise concentrations of chemicals and electrical charges required for different purposes in different parts of our bodies.

Other species, by the way, are much more diverse than humans. We humans each have twenty-three chromosomes, and if the number is different, that's unusual and probably very bad for the person. But some plants are so varied that a member of a species growing at one altitude may have half the number of chromosomes as another individual in the species growing at a different altitude. Compared to other species, humans are monotonously uniform.

Even our closest relatives in the tree of life, chimpanzees, have a lot more variation among them. Their DNA differs from human DNA by only about 1.2 percent. That may not sound like a lot, but it comes to about thirty-five million different bases, five million places where some bases are inserted or deleted, where the chromosomes are somewhat rearranged. Moving a little further away in the family tree, gorillas are only about 1.4 percent different, and orangutans are about 2.4 percent different.

It sounds like a joke: how is a banana like a human? Someone, however, took it seriously and answered the question: bananas are very unlike humans except that they and we both have cells, both grow and power our cells in the same way, and although we may have very different proteins that do very different things, humans and bananas make proteins in the same way. The DNA sequences in humans and bananas are about 90 percent the same.

Maybe that's part of the answer to how our characters can have superpowers. Maybe Papa Suresh was wrong—maybe it's not a case of evolution enabling latent parts of the human genome. Given how many of our characters have ties to the Company, maybe our superpowered humans are transgenic experiments. And a few of the superpowers might possibly, someday, be within reach of people whose DNA included genes transferred from another source.

I'm not suggesting that one of our protagonists has banana genes. But what if Dale Smither (the short-lived mechanic with super hearing from the episode "Unexpected") had ears a little more like a bat than a standard human? What if, as a developing embryo, the cells that would create Niki Sanders's muscles came from an animal with more efficient muscle fibers? Would that account for her super strength? Maybe the Company does more than inject a tracer into people when they are abducted, maybe they're doing some sort of genetic transfer? We know how to insert genes into other organisms—supermarkets today sell tomatoes with a salmon gene inserted into them to help keep them from freezing. And transgenic mice are a standard research tool.

The weird world of chimeras

We do have the technology to insert DNA from one organism into another. In the real world, experimenting on humans the way I just speculated about is obviously unethical and illegal in most or all the countries of the world. (Rules for experiments on humans require that some major benefit offset the risks as well as requiring that the people involved understand the risks and agree to them. While rules concerning animals don't have to meet the same requirements, most organizations still require that the researcher explain why animals must be used, what benefit can result from the experiments, and what will happen to the animals at the end of testing.)

Sometimes, one animal or plant has more than one type of DNA. Although we assume that all of our DNA—barring mutations and accidents—is the same, that's not always true.

Some people and animals and plants are naturally occurring chimeras.

A mythical chimera is a monster made of bits of other animals: it has a lion's head, a goat's body, a serpent's tail, and it breathes fire. In genetic terms, it's a little less fearsome: a chimera is an individual who has cells that descend from more than just one fusion of an egg and sperm cell. Some of the cells have one type of DNA while other cells have a different type of DNA.

When the different sources of DNA are from the same species, there's no obvious indication that the individual is a chimera. For example, in cattle, twins share a single circulatory system before birth, which leads to an exchange of stem cells. In the resulting animals, some cells in their blood, liver, and lymphatic systems contain the DNA of the other twin. When we talk about human fraternal twins, sharing a circulatory system isn't normal, but there's still some crossover of cells. Maya and Alejandro Herrera may have swapped some cells during gestation. Most mothers also retain some cells from the children they bear.

The strangeness and discomfort starts when the source of the other cells is not a very close relative but an animal of a different species. This doesn't tend to happen in nature. (A mule, by the way, is not a chimera but a hybrid. All the cells in a mule descend from the fusion of a single sperm and single egg donated by a donkey and a horse.)

In some situations animals, or parts of animals, can provide a clear benefit to humans: sometimes people with defective heart valves receive organ transplants from cows or pigs. Technically, this makes them chimeras. And people with diabetes benefit from insulin. The insulin used to be collected from livestock animals. Today, in the United States, it is produced in bacteria. The bacteria have been genetically

QUAILING CHICKENS

When someone has a stroke, what sorts of cells could take over the job of the damaged cells? To answer that question, we need to know what cells in the brain control which hardwired behavior. To answer that question, we need to do some research into brains. And to do that, one researcher used chimeras.

In 1997, neurobiologist Evan Balaban, then of the Neurosciences Institute in San Diego and now at McGill University, published a report showing that he had identified areas of the avian brain that have to do with head movement and birdcalls. His experiments were based on creating chimeras, by replacing some of the cells in the brains of developing chickens with brain tissue from developing quails.

He produced chickens that act mostly like chickens—except that their vocal trills and head movements are distinctly like quails'. This shows a couple things. First, that the transplanted parts of the brain contained the neural circuitry for these behaviors. Second, that complex behaviors can be transferred across species. And third, that chimeras can be powerful tools for learning about biology.

FURTHER READING
Frederic Golden, "Cock-A-Doodle Quail," in *Time*, March. 17, 1997.

engineered to produce human insulin, so in this case, the bacteria are the chimeras.

Some larger animals are also genetically produced chimeras. A line of research mice has human immune systems. These animals have been invaluable for tests of new drugs against the AIDS virus, which does not infect conventional mice.

Researchers at Stanford University have made mice with about 1 percent of their neurons made of cells with human DNA. It's still a mouse brain, with mouse architecture and mouse functioning—but it is also extraordinary in that the human cells follow the biochemical directions that allow them to move and be incorporated into the rest of the mouse brain.

But the ethics here get trickier. On the one hand, we can learn a lot about how human brains develop this way. On the other hand, many people agree with Chandra Suresh, who said in the episode "Six Months Ago": "If the soul exists, scientifically speaking, it exists in the brain." If we create a mouse with human brain tissue, will it acquire a soul? Consciousness? Ability to reason in distinctly human ways? How much human tissue is too much? What if mouse brain tissue started developing humanlike brain structures?

In 2003, Chinese scientists at the Shanghai Second Medical University successfully fused human cells with rabbit eggs. The embryos were reportedly the first human-animal chimeras successfully created. The embryos were destroyed after growing in a lab dish for several days.

The discussion about what is ethical is still raging. It is a question not just for the scientific community but for everyone to consider. Do the potential rewards justify the potential risks of this research? What limitations should be put on research? It cuts to the heart of discussions about what defines humanity and what we value.

MYSTERIOUS MITOCHONDRIA

Or maybe humanity isn't just one thing, that set of twenty-three chromosomes that we're so fond of. Maybe we and most of the more complex animals are the results of different organisms fusing. Here's the smoking gun: our DNA is held in the nucleus of each of our cells. But another structure inside our cells, our mitochondria, has different DNA.

Mitochondria are important to the life of a cell: they produce the bulk of the ATP used to power cell activities, and they regulate cell metabolism. Lynn Margulis, a biologist who is now head of the geosciences department at the University of Massachusetts, championed the now-generally-accepted idea that mitochondria originated as self-sufficient cells without nuclei. (These are usually called prokaryotes. Bacteria are prokaryotes, but every other animal or plant you are ever likely to see is a eukaryote, an organism whose cells contain nuclei.)

The endosymbiotic theory suggests that ATP-producing bacteria were enveloped by early eukaryotes, and that the two life-forms benefited so much from the relationship that they came to depend on each other. The mitochondria provide power—they even split and form more mitochondria in a cell if the power requirements are high. But on the other hand, they benefit from the cellular DNA, which encodes proteins required by the mitochondria. The size, structure, and DNA of mitochondria resemble some types of bacteria, which supports this theory.

In her 1981 work *Symbiosis in Cell Evolution,* Professor Margulis argued that all cells with nuclei originated as communities of interacting bacteria, and she suggests that other structures inside cells are also descendants of bacteria. This emphasis on cooperation is a distinct difference from Darwin's emphasis on natural selection. Many biologists today believe that the reality is someplace between these two.

Inheritance

That's interesting, to be sure, but what does this all mean for our superpowered characters?

It sure looks like the ability to have powers is inherited in the DNA. Consider the Petrelli family. After Claire Bennet discovers her superpowers, she hunts down her biological parents, Nathan Petrelli and Meredith Gordon. Both of them also have superpowers, although utterly unlike Claire's. Her uncle Peter also has a superpower, and Claire's grandmother, Angela Petrelli, shows at least hints of having some superpower, although we don't know what it is yet. We don't know whether Nathan's children have superpowers, but otherwise all the living members of the family seem to be affected by them.

Consider another example: the family of Niki Sanders and D. L. Hawkins: she has super strength while he has the ability to travel through solids. This superpowered couple's child, Micah, has a third, and apparently unrelated, superpower that lets him communicate with and control electronics. Meanwhile, D.L.'s niece, Monica Dawson, has yet another unrelated ability: a sort of eidetic muscle memory.

Could the superpowers be the result of inherited mutations? Marvel's *X-Men* universe assumes that all superpowers come from mutations. What is more problematic is that, in the *Heroes* universe, specific abilities aren't inherited.

The difficulty, from a genetic point of view, is that the superpowers are all different. DNA expresses proteins that perform functions. It seems entirely possible that Nathan's DNA is sufficiently different from Peter's DNA that Nathan inherited the ability-to-fly proteins while Peter inherited the ability-to-absorb-powers proteins. What sort of proteins could allow a person to fly, or hear thoughts or create a flame in the palm of her hand? I don't know enough biology even to guess. But it seems less likely that Claire, who has one parent with ability-to-fly proteins and one parent with ability-to-start-fires proteins, ends up with neither type of protein, and instead expresses ability-to-heal-quickly proteins.

That isn't the way heredity typically works. The only way I can see to make sense of this is if we suppose that many people have the potential (for example, a recessive gene) for any of a whole slew of superpowers. But one particular gene might control whether any of the powers are expressed. Only if that gene is turned on will any superpower genes be expressed. Which genes get expressed is a matter of random chance.

For most of our superpowered characters, once one gene is allowed, all the other potential ones are permanently turned off. Peter Petrelli and Sylar are the exceptions, the wild cards in the deck: they are capable of expressing multiple superpowers if they are in contact with people who have those powers. Among the other characters, the Haitian has two distinct superpowers. Expressing multiple genes when usually only one is expressed at a time is unusual, but not unheard of. Take gender, for instance. In humans, most of us have a definite gender, but about 1 percent of children have some level of ambiguity. (The percentage is still being debated, but the intersex characteristic is not.) In some rare cases, a creature can undergo extreme changes in response to environmental cues. Pupfish and a few other types of fish can change sex if they need to, to maintain a viable balance of sexes.

Maybe the ability to have a superpower can be inherited, but the actual expression changes from person to person. Or perhaps the gene that controls whether superpowers are expressed is not a normal human gene at all. Maybe it's something new. Papa Suresh thought it was cropping up in the human population as a random mutation. I suggest that perhaps all of our superpowered people are transgenic experiments by the Company. Or maybe it's something else: maybe it's the result of a new insertion of DNA from a virus.

Mutations and viruses

A mutation is damage to a chromosome. Most of the time it doesn't matter: part of what our cells do is repair DNA. If the DNA is too badly damaged, cells are programmed to self-destruct. Even if they don't, most mutations don't affect much. Sometimes cumulative damage can cause malfunctions that can result in diseases like cancer, but even that—from a larger evolutionary viewpoint—doesn't make much difference. Only when mutations occur in the germ line, in the eggs and sperm that create offspring, do mutations matter to the species as a whole.

On occasion, viruses can do that. They have done that. Viruses are pretty much just strands of DNA (or RNA) in a protective envelope. Portions of human DNA appear to have originated as viruses that managed to insert themselves into the human germ line long ago.

How do we know this? It's one of the fruits of the Human Genome Project. The U.S. Human Genome Project, which ran from 1990 to 2003, determined the sequences of the base pairs that make up human DNA. It also created techniques for analyzing data and at least began to address the ethical, legal, and social issues that arise from knowledge about our genomes.

In humans, genes (the portions of DNA that code for specific proteins) are only about 5 percent of the entire genome. The role of the other 95 percent is mysterious. As mentioned above, parts of our DNA are recognizable as the remains of viruses. Other parts may have regulatory functions that we haven't figured out yet. Some parts of our DNA might function as spacers, making transcription work more easily.

Or, long sections of base-pairs might provide a reservoir of sequences from which potentially advantageous new genes can emerge. And that sounds like the mechanism that Papa Suresh proposed. It is possible: in 2006, Professor Christina Cheng at the University of Illinois showed that nonprotein-coding DNA in cod gave rise to an "antifreeze" protein that helped cod keep from freezing to death. "This appears to be a new mechanism for the evolution of a gene from non-coding DNA," says Professor Cheng; "3.5 billion years of evolution of life has produced many coding genes and conventional thinking assumes that new genes must come from pre-existing ones because the probability of a random stretch of DNA somehow becoming a functional gene is very low…" But it is not, apparently, impossible.

One offspring of the Human Genome Project is proteomics, which is a four-dollar word that means "the study of proteins." The information that's encoded into our genome—at least, the information we know how to decode at this point—concerns instructions on how to make different proteins. Most of our bodies are made of proteins. The biochemical reactions necessary to life are based on enzymes, which are a type of protein.

One of the fascinating things about proteins is that their sequence of amino acids isn't the only important information we need to understand what they do. Their function also depends on their shape. Change the way a protein is folded, and it can do different things. So proteomics isn't just about studying chemistry—it's about structure, too.

Conclusion

The genome we have is all we can work with. The genes in our DNA make RNA. RNA in turn makes proteins. Our DNA could perhaps encode proteins that cause us to grow wings, and with large enough wings we might be able to fly. But a change to our biochemistry subtle enough to not change our standard anatomy isn't likely to result in a built-in jet pack, à la Nathan Petrelli. That's the same problem with Hiro's teleportation and Isaac's clairvoyance: no matter how marvelous the human body is (and it is!), the results still have to obey the physical laws of the universe. We have to finesse the situation, looking for loopholes and areas where we still don't have a good grasp of how the universe works, in order to find any sort of realistic possibility for these powers.

Some of the other superpowers are a little closer to known physiology: a mutation could create a different form for our ears that increases the sensitivity or range of our hearing, like Dale Smither briefly demonstrated (although supersensitive hearing may well be a disadvantage in most urban environments with their incessant aural smog). Similarly, the healing that Claire demonstrates is really just a more extreme and faster version of what people already do, so it isn't utterly unrealistic.

And since we still don't really know what thoughts are, we can't say that it's impossible that a mutation could provide someone like Matt with the ability to "hear" thoughts from other people.

Still, if current research has taught us anything, it is that the mechanisms of life are complex, varied, and messy. There's so much more we need to understand.

Further reading

The National Human Genome Research Institute maintains a Web site that contains a lot of information about genes, genomics, and current research: http://www.genome.org/.

The journal *Nature* maintains a special feature page on the Web, celebrating the fiftieth anniversary of the discovery of DNA: http://www.nature.com/nature/dna50/index.html.

Rick Weiss, "Of Mice, Men and In-between: Scientists Debate Blending of Human, Animal forms," in the *Washington Post,* November 20, 2004, p. A01.

Matt Ridley, *Genome: The Autobiography of a Species in 23 Chapters,* Harper Perennial, 2006.

Bruce Alberts, Alexander Johnson, Julian Lewis, Martin Raff, Keith Roberts, Peter Walter, *Molecular Biology of the Cell*, 4th edition, Garland, 2002. This book is huge, but the sections are also remarkably readable.

Chapter 3

HIRO. TIME KEEPS ON TICKING

We know how to move through time and space—all of us have been doing it since we were born. But think for a moment about what time and space are. It was a comedian who first said, "Time is what keeps everything from happening at once," but it's not a bad start at a definition of time. Similarly, space is what keeps everything from happening right here.

We are all prisoners of space and time. The shortest distance between two points is a straight line—there is no shorter cut than that. And like prisoners serving out their sentence, none of us can make our workweeks any shorter, or make our holidays come any earlier, or last any longer.

Unless, of course, your name is Hiro Nakamura.

Hiro can twist the space-time continuum around his little finger—traveling instantaneously through space and through time.

From one point of view, teleporting is just another method of traveling—although it is a very nifty method that entirely avoids sore feet or airport security. But traveling forward and backward through time is a much weirder idea. Time travel is so far beyond our capabilities that it's a wonder we even have the concept of it. So, for the moment (ha!), let's start by considering time, and what time travel means.

Time keeps on slipping, slipping, slipping into the future—and we all go with it, moving from the past toward the future at a speed of one second per second, measuring time's passage with stopwatches, seasons, and calendars, treasuring or squandering the precious and irreplaceable seconds between the cradle and the grave, but unable to change our rate of travel or our direction through time.

Hiro, unlike the rest of us, can skip wildly across time, into the future—into a time that is not, in the end, his inescapable future—and back into the distant past. We might have no paddle or rudder in the river of time, but Hiro has a powerful outboard motor and is learning how to use it. He sweated and focused his attention first to slow down the clock's hand, but since that first day in Tokyo, stopping time or moving to a different time seems to require no more effort from him than a scrunched-up face, some concentration, and confidence.

Visiting the past is a great device for telling stories, whether it's Marty McFly DeLoreaning back a few decades to deliver a backbone to his future dad, or the *Enterprise* slingshotting around the sun so that Kirk and Spock can visit Earth. Traveling into the far future is also popular, starting with H. G. Wells's *The Time Machine* and continuing on in any number of science-fiction stories and movies. It takes characters away from their homes and puts them into new situations. Stories about time travel are the ultimate road movies, with all of the past and future as the road.

So there are perfectly good reasons why a story might add a time traveler to its mix. But from a scientific point of view, Hiro's ability to travel through time at a rate and direction other than the normal one has profound implications. If he can travel into the past and the future, that would prove that time travel—in the sense that most stories use it—is possible. If time travel is possible, then we must profoundly change the way we think about time, and the universe, and how both of them work.

But what about a less-exciting-than-the-movies version of time travel? Is it possible to travel through time at a different rate than somebody else? This is less exciting, but it would be measurable and reproducible. Let's lower our sights a little. Instead of traveling forward in time a few weeks, what if Hiro just wanted to jump a day ahead? Or an hour? Or a second? How about a millisecond? Is this sort of time travel possible? By that, I mean real, objective time travel, not the "I'm so bored I think time must have stopped" or the "time flies when you're having fun" subjective impression of time.

The short answer is yes. The long answers all start with "yes, *but…*"

For example, "Yes, but most of the schemes that could bring us time travel require the use of black holes and/or ridiculous amounts of power." And here's another proviso: "Yes, but only into the future, not into the past."

Part of the problem is that we don't know a lot about what, exactly, time is. Just as it's hard to get a good idea of what the outside of a building looks like if you are stuck inside it, we have to figure out what time is while we're living inside it.

What is time? We'd better define our terms a little more precisely, even if we have to use a process of elimination to figure out what time is *not*. As near as we can tell, time is not a particle, or a field. It is sometimes considered a fourth dimension. When we move through space, we have the freedom to move in any of three dimensions: forward or backward, left or right, up or down. Plus, we move through time. But time appears to have only one dimension; it's like a line on which we travel from the past to the present and on into the future.

Our passage through time is a little like a squirrel climbing on telephone or power cables. But unlike the squirrel, we can't turn around and go back the way we came. We are constrained to continue moving forward. In the normal course of our lives, we move forward at a steady rate. Compared to our freedom in the three dimensions of space, this is enormously restricting: imagine if you had to live your entire life traveling from east to west, without deviating to the north or south, without going up or down hills, and without any stops. (Actually, I've been on road trips that sometimes felt like that.) That's what time is like.

Still, though, one can ask a physicist, "What is time?" and receive the perfectly serious answer: "The thing that clocks measure." More often, we use time to measure other things, including our lives.

No matter how bad a day you have, there's always tomorrow. We've developed all sorts of timekeeping technologies—from Mayan seventeen-year-long calendars to atomic clocks—to keep us informed about exactly where we are in the trip from Now to Then. "Time and tide wait for no man."

Time and tide, actually, have a lot to do with each other. The ocean's tides are caused by the gravity of the moon pulling at water on Earth. Gravity, it turns out, also pulls at time. (Gravity also warps space, but we'll get to that in a bit.) Albert Einstein figured that out. And, to talk in any more detail about time travel, we need to know a little bit about his General Theory of Relativity.

Einstein time

The problem with the way that we usually think about time (and space) is that, although our mental model works admirably for explaining and predicting human-scaled events, it breaks down at larger scales. The math that nineteenth-century scientists used to describe Earth's orbit around the sun worked great. But when they tried to apply it to the orbit of some other planets, it didn't quite fit. When experimental data doesn't fit a theory, first you check the data. If the data are both good and consistent, then your theory needs revising. That's exactly what Einstein was doing—the starting point for his Theories of Relativity lay in trying to explain some anomalies, some data that didn't fit, in observations of the orbit of Mercury.

EVERYTHING IS RELATIVE

Einstein's theories changed the way we think about the universe. If you know only one equation that has to do with physics, I bet it's $E = mc^2$. (E stands for energy, and m is mass, and c is the speed of light.) One of the implications of this equation is that energy and mass can be converted into each other. This is bizarre. We tend to understand chemical conversions such as when a chunk of wood in a campfire turns into an equal mass of smoke and ash: it gives off heat in the process, but the number of atoms in the smoke and ash is the same as the number of atoms in the wood (and air) at the beginning of the burning. The process of burning, combustion, rearranges the atoms into different configurations, but it doesn't destroy any of them. Similarly, a pound of flour and a pound of butter and a pound of eggs and some dried fruit, plus some effort from a baker and the heat from the oven, combine to form a fruitcake that's a little more than three pounds of either a delight or a doorstop (depending on your views regarding fruitcakes). Matter is transformed, but it is neither created nor destroyed in the kitchen.

But $E = mc^2$ means that mass can *cease to exist* entirely, and turn completely into energy—it is a mind-blowing idea. And it works the other way, too: given enough energy, matter can spring from nowhere. It doesn't happen in our everyday lives, but it is possible. It also means that conservation of energy and conservation of matter (two stalwart physical laws that we used to think were inviolable) are, in fact, not quite right. Instead, we have one law, of the conservation of energy-and-matter.

For our immediate purposes, though, Einstein's theory also means that other things that we thought were absolute...really aren't. Einstein also found that there is no favored frame of reference—no metaphorical place for Atlas to stand to hold up the world: all frames of reference (that are not accelerating) are equally valid.

In our everyday life, space is uniform and unchanging. But at cosmic scales, space is not uniform: it bends, twists, expands, and even appears to have edges. That'll be useful to remember when we talk about teleportation.

Time contracts and dilates

So, all sorts of things that we tend to think of as absolute, unvarying properties really aren't: mass, energy, and space can all change. (Not that there aren't some strict rules about how and when they change—there are.) And one more thing: time is not absolute.

Time isn't a universal invariant, no matter what Dana Scully says. (When in doubt, believe Einstein rather than Scully, or rather, the writers of *The X-Files*.) You can't talk about this moment being "now" here and everywhere across the universe. The farther away you move in space, the harder it is to define "now" at the same time in two places. Also: *the length of one second for me may not be the length of one second for you.* And therein hangs the tale of the twin paradox.

Imagine a set of twins. One twin stays at home. The other one boards a rocket ship and goes zooming off into space, accelerating to a considerable fraction of the speed of light, until she reaches a distant planet. Then she turns around and comes back home. When the astronaut is reunited with her twin, she finds that what was only, say, five years for her was twenty years for her twin. She is no longer the same age as her twin.

So how does that work? Is it just a cosmic sleight of hand? No, in fact it's a simple (although counterintuitive) example of time dilation. Simply put: time dilates (that is, stretches) at faster speeds. The faster you go, approaching the speed of light, the more that time slows compared to someone ambling along more slowly.

In this case, travel really, literally, does keep one young!

(On the other hand, zooming around also makes you heavier, so as your speed approaches the speed of light, your mass would increase, approaching infinity—as though your suitcases weren't heavy enough already! I'll talk more about this in a bit.)

The effect isn't noticeable to human senses at easily traveled human speeds. But it is real, and a real (albeit very limited), feasible form of time travel. So yes, time travel is possible. Not only that, but it has been demonstrated.

CLOCKS IN THE SKY CONFIRM EINSTEIN

We don't notice time dilation in our daily lives, but we have clocks so sensitive that they should be able to notice it. And the test of any theory is the inability of experiments to contradict it. Which is exactly why, in 1971, J. C. Hafele and Richard E. Keating took four atomic clocks aboard commercial airplanes and flew around the world, starting at the U.S. Naval Observatory, where the United States' "official" time is kept. (Actually, they took two trips, going first eastward around the globe, then westward.) Then they compared the time on their clocks against the official clock.

The results confirmed the predictions: traveling clocks experienced less time than stationary ones.

Actually, it's a little more complicated than that. Relativity predicts two time-changing effects: first, that time will slow down as the speed

. . . continued from pg 61

increases. There's also a second effect: that time will speed up far-ther away from the center of a gravity well. (I phrased that in a very Earth-centric way. More properly: time slows in a gravity well. On a universe-wide scale, gravity wells are the exception, rather than the rule.) A clock up at airliner cruising altitude is subject to less gravity, and thus will run faster.

Since 1971, plenty of other people have measured time dilation. They were able to do this even though the maximum speed achieved by human technology is less than 1 percent of the speed of light.

Nevertheless, the twin effects that speed and gravity have on time must be accounted for in a system that many of us use regularly: the Global Positioning System. Because the clocks on the GPS satellites are orbiting so high above Earth, they weigh less, and thus run faster than clocks on the ground—by about forty-six-billionths of a second per day. The speed at which they're whizzing around the earth, however, means that they'll run slower, by about seven-billionths of a second per day. The discrepancy is thirty-eight-billionths of a second, or thirty-eight microsec-onds, per day. The rates of the clocks are adjusted before the satellites are launched, to provide the correct time to receivers on Earth.

How to build a time machine

Great, so we know how to manipulate time's passage, at least a little. But that's a long way from, say, thinking your way back to fifteenth-century Japan. In *Doctor Who*, our favorite Time Lord travels the universe, moving forward and backward in time and space in his TARDIS through something called the Time Vortex. And in Madeleine L'Engle's *A Wrinkle in Time*, the Murry family travels by means of tesseracts, a way of folding the fabric of space and time, allowing them to step to different planets.

The latter, actually, has slightly more to do with reality than the former. Space can be bent, so maybe it is possible to fold it? Black holes certainly warp space and time. At one time black holes were considered possible mechanisms for time travel, but a black hole exerts such incredible gravitational force that anyone who enters its influence would first be "spaghettified" (in the words of physicist Paul Davies), then caught and held. For them, time would slow to a stop, and there they would stay until time ends for the rest of the universe. Not a pleasant thought.

Another, less violent, possibility for time travel (and teleporting) involves a slightly different idea: "wormholes." Imagine an apple: you see one hole in the skin where a worm ate its way in and another hole where it ate its way out. You know that there is a tunnel someplace inside the apple that joins the two holes, but you don't know where the tunnel is. That is pretty much the idea of a wormhole in space. Like a black hole, a wormhole warps time and space. But unlike a black hole, which is a one-way trip to the end of time, a wormhole has (at least two) openings. These ends are connected in a theoretically plausible way, but not through normal physical space.

If wormholes are real, then they exist on an extremely tiny

scale, much much smaller than electrons. If they exist, then they are constantly appearing and disappearing. Physicists have considered whether it is possible to stabilize a wormhole, enlarge it, and send anything or anyone through it. If wormholes exist, and if they could be controlled, the best guess is that controlling them would require something called exotic matter or negative energy. We don't see any reason why it absolutely can't be done—although it would almost certainly require huge amounts of energy—but we're also a very long way from experimental verification of the existence of wormholes. We don't have instruments that can see or manipulate things that small. Then again, half a century ago, we didn't have the tools for imaging or manipulating nanotechnology, either.

If a wormhole could be captured and tamed, then we would, presumably, be able to move either end. Now—remember the twin paradox, which was one example of time travel? Apply that to the ends of the wormhole. Instead of sending a twin off into space, send one end of the wormhole off to, say, Mars. We could use that as a fast method of transport between Earth and Mars. If that were possible, you could step into one end of the wormhole conveniently located at, say, Manhattan's Really Grand Central Station, and complete your step by placing your foot on Martian soil.

But that's not all it's good for. As it speeds off toward the red planet, the traveling wormhole end would also lose some time (just like the traveling twin who sped off in a spaceship), compared to the stay-at-home end. Now bring that end back to Earth. The traveling end has experienced less time than the stay-at-home end—just for example, let's say that the difference is five minutes. If someone steps into the stay-at-home end, and steps out of the traveling end, she will have traveled backward in time by five minutes. For those five minutes, she exists in two places at once. Or, a person could step into the traveling end and out of the stay-at-home end, thus traveling forward in time. For five minutes, she doesn't exist anywhere.

This would be one way to get around the "no travel into the past" proviso—but one could only travel as far back into the past as when the first wormhole was created.

The invariant speed of light

When we considered General Relativity, a lot of forces and properties that we usually assume are fixed turned out to really be flexible. If space can vary, and time can vary, and even energy and matter can change into each other, then what's left that is a universal invariant? What can we depend on? What doesn't change?

The speed of light. No matter what your frame of reference, the speed of light in a vacuum stays the same: just a hair less than 3×10^8 meters per second. Or to think about it another way: in outer space, light travels about one foot every billionth of a second.

VERY COLD, VERY SLOW (BOSE-EINSTEIN CONDENSATES)

When we talk about the speed of light, we almost always mean "the speed of light in a vacuum." Once light enters the atmosphere, or water, or glass, or anything other than vacuum, really, it slows down a little. And occasionally, for very short lengths of time, and in very unusual conditions, the speed of light can slow a lot, to a snail's pace.

In 1995, Eric Cornell and Carl Wieman created a new form of matter, called a Bose-Einstein condensate, using a gas of rubidium atoms cooled to 170 nanokelvin, which is so near to absolute zero as to make no difference to most of us. (A few years later, they won a Nobel Prize for this accomplishment.) Because the condensates are so cold, they have very little potential energy.

Einstein and Satyendra Bose predicted back in the 1920s that cooling certain types of atoms to a very low temperature would cause them to "condense" into the lowest accessible quantum state.

Let me back up for a minute to talk about quantum mechanics. When you get really really small, down in the quantum world of atoms and smaller things, energy starts getting granular. As an analogy, think about sugar: if you're working on a human scale, say, fixing a cup of coffee, you can measure out one teaspoon of sugar, or a little less, or a heaping teaspoon with no problem. So far as you're concerned, you can have any amount of sugar—it is continuous. But say you were fixing a tiny cup of coffee for a gnat. Ms. Gnat wants her coffee a little sweet, but not a lot. At that scale, you're reduced (ha!) to separating out individual grains of sugar. Does she want one grain or two? You (and she) don't have the option of saying she'd like three-quarters of a grain: you (and she) are limited to whole numbers. That's what "a quantum" means: it's an indivisible amount of something.

When you look closely enough at sufficiently tiny things, like an atom, then you can see their quantum nature. Atoms can exist at some energies, but not others. These are called quantum states. Usually, if you have a bunch of atoms together, the quantum states vary, and form what looks (from afar) like a continuous variation.

So, getting back to Drs. Bose and Einstein and Cornell and Wieman: at very low temperatures in a condensate, all those rubidium atoms slow down and each of the atoms in the cloud settles into the lowest energy state it can reach, Not only that, but all the atoms in the entire cloud share the *same* quantum state: in a way, the entire cloud acts as though it were a single atom. The cooling effectively magnifies a quantum effect, which usually only occurs on the scale of single atoms, to a scale big enough for us to see.

Of the many properties of this type of matter, one of the most bizarre and interesting is that it slows down light a lot more than other materials. In fact, in subsequent years, researchers were able to slow light to a crawl. Bose-Einstein condensates could, perhaps, be used as a way to store information encoded onto light, as a sort of a memory storage that uses light rather than hard disks and flash drives for electronic data storage.

And nothing—no thing, no information, no physical influence—can go faster than the speed of light. Remember $E = mc^2$? As you try to accelerate a mass to relativistic speeds (that is, speeds that are a respectable fraction of the speed of light), *the object gets heavier.* Actually, to be more correct about it: its mass increases.

As an object with mass speeds up, it gets harder and harder to push. The energy that you put into pushing it turns into more and more mass. And by the time you get to c, the speed of light, the mass would be infinite.

This is a problem, it turns out, for teleporting.

It is also a problem for writers of science fiction. All those spaceships from *Star Trek* to *Star Wars* that can zoom from star to star in a few minutes or days, cannot travel that fast merely by increasing the speed of the ship. It takes light from our nearest neighboring star (Alpha Centauri) about four years to reach us. For Captain Kirk to travel between stars within a week requires faster-than-light travel. Hence all the talk about warp drives and hyperspace and star gates. For the sake of the story, the storytellers wave a hand

WEIGHT ISN'T THE SAME AS MASS

Weight is the effect of gravity on mass. Since the biggest thing near us is the mass of the earth, we tend to think of mass and our weight on Earth as the same thing. But all mass creates a field of gravity around it. (Even you. Even this book.) That's why the moon orbits the earth and the planets orbit the sun.

Different masses exert different gravitational forces. If you weigh 150 pounds standing on the surface of the earth, you only weigh roughly 25 pounds on the surface of the moon and a little more than 55 pounds on the surface of Mars.

On the other hand, if you could stand on the surface of the sun, you would weigh over four-thousand pounds! First you'd be crushed under your own weight like a cartoon character under a falling safe. Then you'd burn up, adding oxygen, carbon, hydrogen, nitrogen, and minuscule amounts of other elements to the matter on and in the sun.

RIP.

and say, "They are advanced, they know something that we don't know about the laws of physics. We're not going to explain how they can violate the laws of physics as we know them." And through standard use, much like the Greeks with their plays resolved by a deus ex machina, we get used to the idea. We have gotten so used to the idea that people use an acronym to describe it: *FTL* stands for "faster than light."

FTL travel is, effectively, the same thing as Hiro teleporting.

In reality, teleporting and time travel are closely related. Indeed, they are also pretty much the same thing. Einstein's theory of relativity changed the way we think about space and time—developing the idea of "space-time." Just as speed can change the passage of time, massive objects noticeably warp both space and time. When we discuss space-time as a single entity, then teleporting and time travel become different manifestations of the same thing.

Conservation laws

But we've spent a long time out on the edges of physics. Let's visit with some of the more mundane difficulties of time travel and teleportation before we delve back into modern physics.

Both time travel and teleporting run afoul of some conservation laws. If Hiro disappears from Tokyo and reappears in some other time (say, two weeks into the future) or some other place (say, New York City), then what happens to his mass back in Tokyo?

At human scales, things and people don't just disappear, no matter how frustrating it may be when you can't find your keys. (Actually, there is a very, very slight possibility that your keys could just vanish if most of the atoms that made up your keys were to vanish all at once. But the odds of that happening are astronomically low. It is likely that the sun would burn out first.) Even at a quantum scale, a particle whose position you can't determine will, at least, have a definite momentum. (Measuring both at once to less than a certain accuracy is the problem. You run straight onto the horns of Heisenberg's uncertainty principle, which we'll get to in a minute.)

The mass could turn into energy, courtesy of $E = mc^2$. Because c is such a big number, and c^2 is even bigger, we can see that a tiny bit of mass generates a whole lot of energy. If all of Hiro's mass turns into undirected energy when he leaves Tokyo, it would wreak havoc on the place he leaves behind. And by havoc, think about something on the level of a nuclear bomb blowing up. If Hiro teleported like this, then *he,* not Peter or Ted or Sylar, would be the exploding man that causes a huge explosion that devastates a city—and it would happen in Tokyo when he first teleports, rather than in New York City after he arrives.

Suppose we wave a vague hand at that, invoke Einstein, and say, "Hiro somehow converts his mass into energy and somehow uses that energy to propel himself across space and time." Time travel requires massive amounts of energy, if it's possible at all, so fine. But it still doesn't solve the problems with conservation of mass and energy: if Hiro converts his mass to energy and then uses that energy to travel, where does he get his mass on arrival? He wouldn't still be himself, wouldn't be able to yell a triumphant "Yatta!" when he arrives in future New York, unless all the *zillions* of atoms in his body are appropriated from somewhere else, and then all arranged just as they were when he disappeared from Tokyo—an energy-intensive process.

Hiro's mass *could* come from energy converting into matter, but where is that energy going to be taken from? If it's taken from his surroundings, then it won't cause an explosion, but it would still be enormously destructive. If he sucks the kinetic energy from moving objects, they'll suddenly stop moving. If he grabs the energy in hot and warm molecules, then the air and liquids and solids will suddenly freeze. If he takes the chemical energy stored in molecular bonds that hold solids together, they'll fall apart: imagine all the cars and people and walls of the surrounding buildings suddenly turning to dust.

Reconstituting his mass is not the only reason why teleporting would require a lot of energy, although it's by far the biggest one. Consider also the energy required to change Hiro's altitude: the energy required to lift him against the pull of gravity. Both Tokyo and New York are, effectively, at sea level, so that first jump doesn't introduce a big difference. And when Hiro teleports home from future Isaac's loft

REALLY BIG NUMBERS

"Zillions" is hardly a scientifically useful number. When we talk about the number of atoms in Hiro's body, how big a number, what orders of magnitude, are we talking about? There are roughly ten quadrillion (10^{16}) atoms in a medium-size human cell. And there are a trillion (10^{12}) or so cells in a normal human. Alas, English lacks a good word for 10^{28} of anything.

On the other hand, English has a perfectly good word for something with roughly 10^{23} atoms. Avogadro's number, which describes the number of atoms in one *mole* of an element or molecule, is 6.0225×10^{23}.

The idea of a mole was introduced because for chemical reactions, it's often important to have the same number of atoms (or molecules). If you have some hydrogen gas and some oxygen gas and you'd like to make water molecules, then you need two hydrogen atoms for every oxygen atom. Hydrogen atoms weigh a lot less than oxygen atoms, so weight is an awkward way to keep track of the number of atoms. And gases are compressible, so volume isn't a reliable method for measuring amounts either. But if you take two moles of hydrogen, and one mole of oxygen, then you'll have the right proportions to make water efficiently.

high in a building to the Tokyo subway, he's gained as much energy as if he'd fallen the entire way. Later on, in the episode "Parasite," Hiro teleports himself and Ando (and a sword) from Linderman's vault in Las Vegas to the Deveaux building five years after the explosion.

Las Vegas is over two thousand feet above sea level. The Deveaux building can't possibly be two hundred stories tall. Where did the extra energy go? Could the need to balance out an energy equation be part of the reason that Hiro sometimes ends up in a different time when he's only trying to teleport? How could changing times be anything less than an energy-intensive endeavor?

Even if Hiro can manage changes to his potential energy, he needs to manipulate his momentum. Isaac Newton said, "Objects in motion tend to stay in motion." Tokyo and New York are two different points on the surface of the globe. We don't tend to feel it, because the earth's gravity is pulling us down, and the atmosphere and objects around us are moving with us, but the truth is that we are on a quickly spinning globe hurtling around the sun. The circumference of the earth at the latitude of Tokyo (thirty-five degrees north of the equator) is about 20,500 miles. At New York's higher latitude (forty degrees north), it is about 19,200 miles. Since both cities make a full rotation each day, that means that when Hiro is standing still in Tokyo, he's actually moving at about 854 miles per hour. When he's standing still in New York, he's actually moving at something more like 800 miles an hour. That extra 54 miles per hour would be enough to slam him into a building as hard as a highway car crash—only without the air bags.

It gets worse, though. Hiro's really big problem with momentum is that the two cities are more or less on the opposite sides of the planet: at any given moment, the two are *spinning in nearly opposite directions.* So when Hiro leaves Tokyo and appears in New York, all of a sudden the earth's gravity is pulling at him from a different direction, and everything around him is moving at about sixteen hundred miles per hour, relative to him. It has the potential to be very gory. (And would undoubtedly confuse the NYPD. The crime scene would look like Hiro had been shot out of a cannon.)

And those calculations are ignoring the fact that the planet is also moving around the sun at about sixty-seven thousand miles per hour. If this were a textbook, I'd say, "The implications of this will be left as an exercise for the reader." But it's not, and we can just conclude that whatever method Hiro has for teleporting had better expend some energy to change his velocity—and somehow do it gently enough to keep from splatting him like a bug on a windshield when he arrives.

This level of complexity seems to be a bit much to ask of Hiro's previously undeveloped talent, without having a teacher or any mechanical assistance.

And what about when Hiro travels from New York's Kirby Plaza to fifteenth-century Japan at cherry-blossom time? Earth may not be—and probably isn't—in the same area in space! And then there's Hiro's body—not just the mass (although there is that) but he's also transported the information carried in his memory, his DNA, the arrangement of his cells, his facial features, his hairstyle... (there's a lot of information implied in even the dumbest person). Information and matter have just appeared out of nowhere! If Hiro can jump backward in time, then we can kiss cause and effect good-bye.

Hiro may be close to running into an ancestor of his. Hiro didn't meet a lot of people in Old Japan but it is possible that one of the people he met might have been a many-times-great grandfather or grandmother of his. And it is possible that he killed that ancestor. If so, then he would have changed history, and his particular set of genes would never have occurred in his own time. So by time traveling, he makes it impossible for himself to have ever existed. What a paradox!

The grandfather paradox and Heisenberg uncertainty

Being able to change the past leads to something called the Grandfather Paradox. Suppose that your grandfather, in his latter years, builds a time machine. Suppose you use the machine to go back in time and maybe even meet the man who will eventually become your grandfather. No problem so far. But further suppose that, for whatever reason, you kill that man before he has a chance to father a child? In that case, he never becomes your grandfather and he never builds a time machine. You don't exist, nor does the time machine, and so you can't have gone back in time to kill him. Paradox!

There are variations on this idea that are less violent, or at least less malicious: what if you go back in time, eat an apple, and spit out the seed in Illinois? The seed grows into a tree that falls on a young Abraham Lincoln, who therefore doesn't become president, and the fate of the United States is changed forever.

We see a classic example of the paradox when Future Hiro travels into the past to tell Peter to "save the cheerleader" in the episode "Collision." New information has been created in the world that changes the future, so that Future Hiro doesn't exist. But he had to exist, because he appeared to Peter...

Does you brain hurt yet? Mine does.

It's a small step from assuming time travel is possible to arguing about free will vs. determinism. Do we, in fact, have free will, or are we predestined to do and become certain things? In modern America, we tend to come down firmly on the side of free will: "I am the captain of my fate, the master of my soul" (William Ernest Henley's poem "Invictus"), although some physicists aren't so sure about that. But combine free will with time travel and again—paradox! If you change the past, you change the future, and then (no matter where you started from) the original future no longer exists.

Hiro, by the way, tries to have it both ways. As he talks Ando into accompanying him to the United States, he first claims that, "By taking action we've changed something"—that they can change the future he has seen. But within minutes, he's also claiming that Ando's actions are predetermined: "The comic book says we fly together, so we fly together." Hiro is both adorable and admirable. But in this case, he is not logically consistent.

Even if Hiro did nothing but observe another time, he might still change it. Remember Heisenberg? He stated his famous uncertainty principle, that we can't determine both the speed and location of a particle to better than a certain

STARING INTO THE PAST

We can, and regularly do, look at the past: we look up into the night sky and see the light generated by stars a few years (in the case of Alpha Centauri) or decades or centuries or millennia ago. The sky that we see doesn't really exist: it is a collection of artifacts, light generated at different times in different places across the galaxy, which just happens to arrive at the surface of the earth at the same moment. Our seeing the light changes the light: the particles of light, or photons, are absorbed in the structures at the back of our eyes, and turned into electrical signals in our neurons.

Merely seeing the light does not, of course, change the stars that generated the light long in the past. If it did, that would imply retrocausality: something that happens in the future causing an effect in the past. That's a downright scary idea, because it would mean that from our vantage point (traveling from the past through the present and into the future) the world would be unpredictable, and not just "fuzzy" at a quantum level as predicted by Heisenberg.

value—and not because we lack tools that are sensitive to measure the value, but because that level of certainty doesn't exist. Heisenberg stated that observing alters the observed. In other words, the mere act of watching transfers some energy, and perturbs the state of the observed thing. He was talking in terms of quantum effects, but when you add in time travel, there's no reason why it wouldn't create bigger effects, too. So, just by observing the future or past, you'd change it, and thus change the world in your own starting point.

Isaac and information from the future

What would happen if retrocausality were real? Could we find statistics to predict previous actions? How far into the future would we need to look to understand why something in the past occurred?

The problem with this idea is that the world *is* predictable, in a general sense. And for Isaac Mendez, the future is literally foreseeable. In order for Isaac to paint the future, he has to envision it in his mind's eye. But for Isaac's precognition to work, this information—the visions that he paints—must travel from the future into the past. Precognition pretty much equals the time travel of information.

Wouldn't his observation be sufficient to change the future? Maybe not enough to change the events that he paints. Or maybe it is enough.

Which leads us to the multiple universe theory

But what if there are an infinite number of similar universes? Every time you "time-traveled," you would also be skipping to a different universe. So: killing your grandfather in one universe doesn't kill him in the universe that you came from.

And, if we're positing infinite universes (as some serious scientists do), then time travel is not really time travel—maybe it is just jumping to a universe where things happen more slowly, where history is delayed, so that one's grandfather is still a young man when one gets there. Maybe when Hiro appears to stop time, he's just moving to a universe where time moves much much slower than he can? That would explain how he visited a New York where a bomb went off, and visited a world where Sylar killed Claire.

The many-universes idea might also explain some oddities—like how come Hiro isn't immediately obvious as a stranger when he wears glasses and today's clothing, and talks in modern Japanese to people who lived four hundred years earlier. And for that matter, would he and an Englishman of that time be able to understand each other? It seems unlikely, in a single consistent universe.

If Hiro can time-travel, then it's possible—so why wouldn't other people travel through time? In which case, where are they? Wouldn't historians and tourists flock to the important events in time, boosting the attendance? Wouldn't some of these people be noticeable because among any group of people some of them will be, honestly, idiots, dressing and talking and acting in ways as inappropriate as *Harry Potter*–style wizards trying to pass as Muggles? Why have none of them been sighted when they appear and disappear?

This isn't proof that time travel doesn't exist, but it does put some commonsense limitations on how it might work. It might mean that time travel would either be rare, or incorporate some sort of camouflage, or not be possible yet, or maybe we live in a universe that somehow repels time travelers.

Mostly, I've been trying to square what we see on *Heroes* with the universe as we know it. Let's let Hiro rest for a moment and consider what is real for us. Given huge cosmic forces and some very clever engineering, it might someday be possible to send a human through time in a controllable nonstandard direction and rate, using wormholes or strings, or some other exotic loophole.

That's a long way away. So is there cause to hope that we might someday have the technology for time travel? Yes. Because there are still some very basic questions about the universe that we have not answered.

What we don't know, yet

We understand the effects of four fundamental forces that bind the universe together: the strong and weak nuclear forces (which work at very small distances), the electromagnetic force (which, in addition to electricity and magnetism, includes radio waves, visible light, and X-rays), and gravity (which works at large and very large distances). We've learned a lot about all these forces in the past 100 to 150 years. The twentieth century, in particular, was great for physicists. General Relativity, as we've discussed, tells us a lot about how the universe works on the scale of planets and larger sizes. It also brought us an understanding of quantum mechanics, which explains how the universe works at sizes of atoms and smaller things.

But the two schema don't meet gracefully. For decades now, the Holy Grail of physicists has been to find a single theory that incorporates both gravity and quantum mechanics—and is experimentally verifiable. After a period when it was fashionable to call these "Theories of Everything" and "Grand Unified Theories" (or GUTs, for short), work that tries to join both the very large and very small is now focusing on "quantum gravity."

One of the proposed theories, superstring theory, suggests that particles are actually many-dimensional strings. At least one string theorist conjectured that if two infinitely long strings were parallel to each other, then time travel would be possible by moving in a figure eight around them. (String theory, by the way, hasn't been experimentally verified. There's some question about how it *could* be experimentally verified. At the moment the theorists are working on making sure that it is at least internally, mathematically, consistent.)

Maybe when we understand more about how the universe works, we'll also learn more about time. And once we know that—who knows how hard it'll be for someone to figure out how to bend the space-time continuum—just like in *Star Trek*! Or Hiro.

Further reading

Paul Davies, *How to Build a Time Machine*, Viking 2001. A bit hand-wavy, as it must be without including the math. A good introduction to current theories, the people involved, and the theory of relativity.

Alfred Bester, *The Stars My Destination*. Classic SF story that posits a limited type of teleportation and the effects it has on society.

Stephen Hawking, transcript of lecture "Does God Play Dice?" Available at http://www.hawking.org.uk/lectures/dice.html.

Carl Sagan, interview, "Sagan on Time Travel," part of the NOVA

Time Travel site. Available at http://www.pbs.org/wgbh/nova/time/sagan.html.

New Scientist's special report on the Quantum World. Available at http://www.newscientist.com/channel/fundamentals/quantum-world.

Madeleine L'Engle, *A Wrinkle in Time*. Farrar, Straus & Giroux, 1962. (Has been reprinted many times.)

Chapter 4

NATHAN FLIES

Nathan Petrelli flies. That's a classic superpower, dating back to the first superhero. Between flying, his residence in a grand metropolis, and his chiseled jaw, all Nathan needs is a cape to trigger Superman comparisons.

But how, exactly, does he fly?

Does he fly like a bird? A plane? Superman? (Well, obviously he flies like Superman. But can he just jump very high—which was the origin of Superman's flying power, by the way—or is he doing something else?) Or maybe Nathan isn't actually flying, maybe instead he's floating the way Superman in the movies seems to? Maybe somehow he can exempt himself, at least some of the time, from gravity?

Gravity 101

What keeps us, and pigs, from flying? Gravity. The gravity of the earth pulls us toward it, keeping us from falling out into space. The earth's gravity well keeps us stuck to the ground the same way it keeps the moon in orbit around the earth.

FLYING IS JUST FALLING AT THE GROUND BUT MISSING, AKA ORBITS

I really love watching Road Runner and Wile E. Coyote cartoons on Saturday mornings, especially when Coyote straps roller skates to his feet and a rocket to his back and zooms off down the highway. But I always feel he's missing an opportunity: if he could just run fast enough, Coyote should be able to pick up his feet and go into orbit and no longer worry about the rocket running out of fuel. If he misses Road Runner on one pass, he can just relax and wait a few minutes and then try again on the next orbit.

An orbit is a very cool concept. The moon, for example, is in orbit: it is constantly falling toward the earth, but (because it is moving in another direction as well) it is also constantly missing the earth. So do the International Space Station and countless satellites.

But most satellites are up in space. How fast would Coyote have to go to get into orbit at ground level? I am going to make some unrealistic assumptions to make the calculation easier, but considering that we're talking about an apparently immortal supergenius coyote who regularly plays with dynamite and receives mail-ordered tools, I feel confident that our assumptions aren't the most ridiculous thing about this scenario.

First estimations often contain assumptions to make the calculations easier. Everyone knows that they are unrealistic, but for a quick guess or a back-of-the-envelope calculation, that may not matter. Still, assuming things to make the math easier can lead to some pretty funny descriptions. There are lots of engineering jokes that either begin or end with "Assume a spherical chicken." In this case we don't need to assume a spherical Coyote.

First, let's assume that we want an orbit three feet above sea level. But Coyote has this alarming habit of slamming into things at high speed, which would get in the way of demonstrating a decent orbit. So let's also assume that the earth is *very flat,* at least near his orbit: no Mount Everest, no hills, no trees, not even any waves higher than three feet, and definitely no trucks, billboards, or buildings. The molehills had better be pretty small, too.

Third—and this is a big assumption that is utterly incorrect in the real world—let's ignore the presence of air. In realistic terms, the same air that makes a dog's ears flap when it sticks its nose out the car window is a major force in slowing down anything that flies. But for right now we get to ignore it.

The variables involved are the force of Earth's gravity at sea level and the diameter of the planet. (Coyote's weight is *not* a factor. Heavy objects and light objects fall at the same rate. Galileo demonstrated that.) How fast would Coyote have to be moving horizontally so that by the time gravity pulled him down, he'd already have moved around the curve of the earth—so he could keep falling and missing the ground? He'd have to zip along at almost five miles per second. At that speed, Coyote would make a complete orbit around the planet in under two hours. His in-flight movie wouldn't even have finished before he got a chance to grab at the unsuspecting Road Runner for dinner. Road Runner wouldn't be able to hear him coming either, given that he's moving at roughly twenty-three times the speed of sound. Heck, the sonic boom alone might knock over the bird.

. . . continued from pg 89

That's an unrealistic example, but it illustrates some interesting points about more conventional objects in more conventional—that is, much higher—orbits. Objects going too fast don't get caught orbiting a gravity well. Their path may bend some, but they zoom past and continue on their way. Objects whose combination of speed and distance from a gravity well are not large enough to maintain an orbit end up spiraling into the gravity well. This is exactly what happened to Sky Lab in 1979. After years of service, it ended up smeared across the Australian Outback. And there's a particular height, 22,284 miles above the surface of Earth, where a satellite can orbit the earth exactly once a day, so it stays over the same spot on the planet at all times, in a "geosynchronous orbit."

NASA has a really nice graphical explanation of orbits online. See: http://spaceplace.nasa.gov/en/kids/orbits1.shtml.

Gravity (from the Latin *gravitas*) is serious business—one might even say, a grave matter. It is one of the four fundamental forces in the universe. Gravity is not optional. So, if pigs are ever going to fly, it'll have to be in some way that doesn't depend on gravity turning off.

All flying creatures and machines have to generate some sort of lift, some way to push against gravity to get up in the air, and some way of staying up in the air. Short of some sort of antigravity, Nathan—and now West, too—has to expend energy to push himself up into the air.

If he wants to move sideways once he's in the air, he needs some way to push against air drag—this is typically called thrust. The difference between lift and thrust is easy to remember if you recall Nathan taking off in "Hiros." From the ground, he flies straight up, and then seems to pause for a moment. That's lift. After the pause, he zooms off sideways, leaving a smoke ring behind him. That part requires thrust (the smoke ring is optional).

Of course, lots of things fly: birds do it, bees do it, and humans in airplanes and hang gliders and hot-air balloons do it. Why shouldn't Nathan fly using one—or more likely several—of these methods?

Not *those* birds and bees

Animals that fly are built differently from animals that move on top of the ground. Their upper size is limited—with a six-foot wingspan, the American condor is pretty darn big, but it's still no elephant. Or hippopotamus. Or whale. The wingspan of birds is pretty big compared to their total weight.

For example, the bird with the longest wingspan, the wandering albatross, can measure over eleven feet from wing tip to wing tip and might weigh up to 25 pounds. In comparison, the orangutan has the longest arm span among the primates—as long as seven feet from fingertip to fingertip, but it can weigh up to 180 pounds. Compared to the bird, the great ape is just plain stumpy. Those long arms work great for orangutans, who use them for climbing through the jungle. Their arms are much longer, proportionally, than humans'—more of our muscle goes into our legs, all the better to run with.

Bird bodies are also well adapted for flight in other ways. Bats, which are mammals like us, share many of these adaptations for flight. Both bird and bat bones are hollow, which provides strength but less weight. (All bets are off, however, once we start talking about flightless birds like penguins and kiwis and ostriches. Or roadrunners.)

Birds also have remarkably fast metabolisms, which helps when a lot of effort is required. Flying is hard work! Which brings up a point: if Nathan flies like a bird, then he probably has the metabolism of a bird, which means he had better eat like a bird. Sure, we say "eat like a bird," but that really means consuming nearly his own weight in food each day. Does Nathan chow down constantly or use his ability to fly very very sparingly? No wonder that after he escaped from Mr. Bennet and the Haitian in the episode "Hiros," he made a beeline (so to speak) to a diner! He needed fuel!

Which leaves us with Nathan, whose anatomy (presumably) resembles that of a normal human, but who can fly like

THAT BUMBLEBEE STORY

You've probably heard the story: even after an engineer "proves" that a bumblebee can't fly, the little-insect-that-could merrily flutters away, ignoring the best calculations known to man.

On one hand, this story is a useful modern parable for persevering in spite of disapproval or disbelief. It's also useful for pointing out that while scientists study nature, the models we develop to explain it may not be correct.

On the other hand, the bumblebee story can also be used to discount the conclusions of science altogether or to lump well-tested theories in with pseudoscience, the way some creationists try to suggest that evolution and "intelligent design" are equally valid theories.

The darn story is a teeth-gnashing pain in the rear. So far as we know, the story originated in the 1930s. Given what was known about aerodynamics and fluid flow at the time, we can presume that the engineer in question (if he existed) calculated the lift generated by a bumblebee's wings *as if they were fixed like an airplane wing*. He might even have added in the speed of a typical bumblebee's flight. Not surprisingly, such a tiny fixed wing would not be able to keep even the tiny weight of the bumblebee aloft.

The point is that the model was obviously flawed. It may have been a good place to start, but any scientist or engineer, on seeing the results of the calculation and the clear evidence of actual bumblebee flight, would have looked for a more realistic model. Since the 1930s, we've learned a lot about the lift generated by wings that move.

FURTHER READING

Ivars Peterson, "Flight of the Bumblebee," *Science News Online* 166 (September 11, 2004), available online at http://www.sciencenews.org/articles/20040911/mathtrek.asp. More about the bumblebee story from the admirable pen of Ivars Peterson.

Michael Dickinson, "Solving the Mystery of Insect Flight," *Scientific American*, June 17, 2001. More from a researcher who spent years figuring out a valid model for bumblebee (and other insect) flight.

a bird. There are some problems with this idea, besides sheer weight. The most obvious one is: where are his wings? But say we get past that and assume that he has, say, retractable invisible wings built into his arms, we're still left wondering why he doesn't look like an orangutan. Wouldn't he need longer arms to provide wings in proportion to the rest of his body?

While we're at it, how do wings work? Basically, it's all about lift, whether we are talking about birds' wings (which are jointed) or airplane wings (which are fixed). For now, consider how they work when the bird is just gliding. Wings generate lift by changing the air pressure around them. If the wings can make the air pressure below larger than the air pressure above, then the air itself will push them upward— creating lift. All wings manage this by their shape.

Imagine the cross section of a biscotti: it is flat on the bottom and gently curved on the top. Wings are, more or less, shaped like that biscotti. When air flows past that shape, the air above it has to divert farther—has to take a longer path—than the air below it. So there are fewer gas molecules per unit of area right above the wing than there are below the wing. There's more stuff underneath than above, which causes the air pressure to be lower on the top than on the bottom, which in turn provides lift.

It's more complicated than that. We've ignored thrust entirely. The angle of the wing and the trailing edge of the wing alter how air moves around it. But that's the gist of how wings work on airplanes as well as for birds. Birds, though, have some advantages over fixed-wing aircraft. They can beat their wings to create more lift: as the wing comes down, the bird gets lift upward, and as the wing comes up it can swivel and fold (to keep from pushing the bird down).

As mentioned above, wing-flapping creatures fly by manipulating air pressure—that thing that we ignored while

IS NATHAN *IMPULSIVE?*

In the scene with Linderman's goons, he didn't look like he was flying the way a bird would fly. It looked like he took off like a rocket, but Nathan doesn't have rockets strapped to his shins, so how does he zoom into the air? What if he launches simply by jumping, like a turbo-basketball player? He would need some serious leg muscles, exerting impressive force.

If he is jumping up into the air, then he does all his pushing against gravity in the very first moment. That sort of force is called an impulse. And it's different from a rocket, which keeps pushing against gravity as it rises. Nathan is roughly six feet tall, and probably weighs, oh, 180 pounds? Let's assume he goes up the length of a football field, a hundred yards, before he more or less stops. So how fast does he need to go at takeoff in order to get that high? If his highest velocity is when he leaves the ground, and steadily slows until his velocity at a height of a hundred yards is zero, then when he leaves the ground, he has to be going about ninety-five miles per hour!

Those are some calves!

thinking about Wile E. Coyote's orbit. Suppose some astronauts played Noah's Ark and took a variety of animals to the moon, each in its own specially tailored space suit, of course. (The engineering of space suits for animals would be a fascinating challenge, too.) Without an atmosphere, none of the birds would be able to get off the ground—flapping their wings wouldn't avail them at all. The champion for airborne time would probably be a kangaroo or (more likely) a cricket—some creature with ridiculously strong jumping muscles.

Rockets and hang gliders

Using wings and jumping are the most common ways to fly, but not the only ones. Consider the hobbyist's model rocket: a solid-fuel engine hurtles it into the sky (since the exhaust first pushes against the ground, then pushes through the air) then, as the fuel is exhausted, a parachute deploys, letting the rocket float gently back to the ground, where its owner has at least a fighting chance of recovering it.

Since the days when scientists like Robert Goddard and Wernher von Braun were pioneering rocketry, engineers have developed better ways to use controlled explosions to push against gravity. The rest of the vehicle design has been refined and optimized, but even a brick will fly if you strap a rocket to it.

Remember that rocket backpack that James Bond (and, if I remember correctly, Flash Gordon) strapped on? These devices, called jet packs or jet belts, really work. But the flights they provide are noisy, dangerous, short, and expensive. The amount of fuel that even James Bond can carry on his back is sufficient for only a few seconds—less than a minute—of flight.

There are at least three companies that make a limited number of jet packs, but they tend not to sell them to the general public. Your chance of getting killed playing with a jet pack (and the companies' chances of getting sued by your heirs) is far too high.

Look at it this way: you are strapping explosives to your back and expecting them to burn in a steady and controlled rate, rather than go off at once or in ways that'll kill or horribly maim you. Even if the fuel burns in the nozzle or nozzles as engineered, this thing is strapped to your back with very hot gases shooting out the bottom. That is roughly where your legs and feet are. What is to stop your lower half from getting charbroiled?

Assume that intelligent engineering and flawless operation remove your legs from the oven and provide enough thrust to get you up in the air. The human form is admirable, but it is not designed for stable flight, so you're going to have to control the jet without the benefit of stabilizers. But if you're skilled and the jetpack is designed to compensate for a person of variable weight being strapped to one side, you can probably get a good twenty or thirty seconds of thrust out of the jet, enough to get you a few hundred feet up. That's great; it's probably what you wanted when you started this adventure. It's great, at least, until the fuel runs out and you are left with all the aerodynamic properties of an unbalanced brick. You might possibly be high enough for a parachute to deploy before you hit the ground, if only you can keep from tumbling. It's no wonder that jet packs that are advertised for sale have price tags of hundreds of thousands of dollars!

If you were to use a jet pack, it would be wiser to stay fairly low and take only short hops. But heck, if you're just going to do that, you could do it a lot cheaper using a different type of flying "personal air vehicle." One company, Trek Aerospace, traded off the portability of the jet pack—and, indeed, the jets—for longer flight times. A rotary engine that runs on airplane fuel powers two counterrotating fans, mounted on an eight-foot-tall frame. The pilot stands on the front of the frame, the engine is on the back, and the fans are above and to the sides of the frame, making it nine feet wide. This contraption is a bit bulky for Mr. Bond, but still pretty neat-looking. It's not for sale, but it is being marketed.

Nathan—especially when threatened by Mr. Bennet and the Haitian, acting as Linderman's goons—took off like a rocket. But just as he lacks wings, there's really nowhere that he could have stowed rockets, given that he was wearing only his boxer shorts at the time. (Okay, those of you with dirty minds can stop sniggering now.)

The floating world

Hang gliders are at the other end of the spectrum from rockets. People who use hang gliders tend to climb to high places and jump off, since the contraptions are basically a big unpowered (and blissfully quiet) wing. It is true that if a hang-glider fan can find herself a current of air that is moving upward, then she can gain height. But in general, hang gliders don't so much have a way of flying as they have a slow (and enjoyable) method of falling.

Hot-air balloons aren't as dramatic as rockets, but they are particularly cool because they use the neat trick of making gravity do the work of pushing them upward. Gravity pulls down solid objects (like this book), and liquids (like the ocean), and also gases (like the atmosphere). If you drop a penny in a glass of water, the penny falls to the bottom because it is denser than the water. So a penny's worth of water is displaced. When the barista at your local coffee shop makes you a cappuccino, a similar effect keeps the foam on the top of the cup, because the liquid is denser than the foam. That's the same reason why cream rises to the top of milk and why oil floats on top of water, causing oil salad dressings to separate. (It also causes oil slicks in the ocean, but that wouldn't fit in with the other food examples in this paragraph.)

The point is that by introducing a less dense structure into higher-density structures, one can create lift. Solids are usually stiff enough to keep from sorting themselves by density (it would be inconvenient, to say the least, if your metal paperweight went through your wooden desk like a penny through water), but liquids and gases do it all the time.

In the ocean, there are fish that contain a particular structure, called a cuttlebone, full of gas, that works to buoy up the animal, allowing it to swim with less effort. (We'll revisit cuttlefish in more detail in Chapter 6.)

Here's another example: we are creatures made largely of slightly salty water. Salt water is denser than fresh water. We can float in freshwater swimming pools. We can float even more easily in salt water. And in the Dead Sea, where the salinity is very high, we float most easily of all. If Nathan were somehow less dense than the air around him, he could float, almost like we do in very salty water.

Buoyant balloons contain less dense gases rather than less dense liquids: the weight of a hot-air balloon's fabric and basket and pilots (their mass pulled down by gravity) is offset by the buoyancy of the gas inside the balloon. It's not that the gas inside has any sort of antigravity—it's that the atmosphere outside the balloon is pulled down by gravity more. Hot air is less dense (that is, has less mass in the same volume) than cold air, which is why hot air rises. So when the pilot fires the burner, she heats air, which fills the balloon, which becomes an area of lower-density air than the surroundings.

Children's helium balloons float for the same reason: the

helium has less mass than the nitrogen-rich atmosphere that we breathe, so its weight is less than the air around it. How heavy is air? Air pressure depends partly on the weather. But in general, the closer to the center of the earth, the more a gas is compressed, so the denser it is, the higher the pressure. At sea level, air pushes down (and sideways, and up from below—it's the pressure we're talking about, not gravity directly) at almost fifteen pounds per square inch!

Nathan looks like a big solid guy, but maybe he weighs less than we do. Maybe he has hollow bones or a cavity filled with buoyant gas? This would mean that when he's up in the air, he's not so much flying as *floating,* like fish swimming in the sea, or hot-air balloonists.

Only, if that's true, then why doesn't he have problems staying anchored in a stiff breeze? If he depends on buoyancy to fly, Nathan must have a fast way of generating, holding, and shedding low-pressure gases. Maybe he has something like a swim bladder?

Effects of flight

We know that the force of gravity from Earth's mass decreases as we move away from it (for example, astronauts on the International Space Station are effectively weightless), but does that help Nathan fly? No, not appreciably.

Why not? Because Nathan probably doesn't go nearly high enough to feel the change in weight. Remember, when we fly in commercial airplanes traveling at thirty thousand feet above sea level, we're not floating down the aisle to the lavatory. At this distance, the force of gravity doesn't noticeably decrease. The International Space Station, by contrast, is floating at over a million feet (more commonly expressed as 220 miles).

The atmosphere is thin. Six miles might seem like a short commute to work, but going six miles straight up puts you near the top of the troposphere. Have you seen pictures of climbers on top of Mount Everest? They're typically cold, exhausted, and either sucking on bottled oxygen or panting for breath. And Mount Everest, the highest peak on the planet, is "only" 5.5 miles above sea level.

If Nathan (not to mention West and Peter, and in at least some versions of the future, Sylar) is capable of producing enough lift to get up into the air, and enough thrust to control his movement through the air, he still has to deal with the effects of *being* in the air. Humans evolved to live on the surface of the planet, and even then, there are plenty of parts of the surface that aren't amenable to us: either too cold, too hot, too wet, too vertical to easily walk on, and in some cases, just plain too high. Even the people who live at the foot of Mount Everest, who have adapted to the high altitude and have large lung capacities, stay near the bottom of the mountain.

Air Pressure

At six miles high, the air is thin, and oxygen is thin with it. We're comfortable when we fly at thirty thousand feet, but that's because we're sitting in a pressurized cabin. (The Concorde cruises above sixty thousand feet, yikes!) If Nathan is flying without mechanical help, he's exercising hard, and he's going to need oxygen. And since, so far as we know, Nathan has to breathe just like the rest of us, he's probably flying at considerably less than thirty thousand feet. (Birds, by the way, have occasionally collided with planes at altitudes this high. A vulture collided with a commercial aircraft over Abidjan, Ivory Coast, at an altitude of thirty-seven thousand feet in 1973. And a flock of cranes were sighted at more than twenty-seven thousand feet.)

The troposphere, which extends from the earth's surface to between six and ten miles high, is where weather happens, where commercial airplanes fly. Three-fourths of all the gases surrounding our planet are in the troposphere. Almost all of the rest is in the stratosphere. But humans can only survive without mechanical help in the lower half of the troposphere. People's bodies work well on the surface of the earth, where there's plentiful oxygen in the air and changes in temperature and pressure occur relatively slowly. But as we rise, the air pressure drops and with it goes the amount of oxygen. In other words, there is progressively less air and therefore less oxygen per unit volume as you ascend to higher altitudes.

Therefore each breath of air that you breathe at, for example, fifteen thousand feet above sea level has about half the amount of oxygen of a breath taken at sea level.

Differences in air pressure cause a lot of effects. I live in Boston, at, effectively, sea level. When I send a cake recipe to my friend Val in Denver, which is a mile higher than I am, she has to adjust it before trying it out. Mountain climbers know that water boils at lower temperatures at great heights—it's possible to drink a cup of coffee that's just boiled on the camp stove without scalding your mouth. The lower pressure can also cause altitude sickness.

Many people can breathe just fine up to about twelve thousand feet, and plenty of people visit altitudes that high in the Rocky Mountain National Park, for instance. But the efficiency of our breathing (namely, how much oxygen we get from each breath) decreases with the altitude, with our bodies becoming less and less efficient until we can no longer think clearly or even stay conscious. If he were completely deprived of oxygen, Nathan (or anybody) would die in about eight minutes. But he could find himself in deadly danger much more quickly if he flies too high to breathe sufficient oxygen. Hypoxia—the lack of sufficient oxygen—can be a killer simply because when you are subject to it, you can't think clearly, and may not realize that there is any problem at all. And breathing quickly (hyperventilating) isn't a solution either: that just decreases the amount of carbon dioxide in your blood, which also creates problems.

The effects of hypoxia include euphoria, reduced vision, confusion, inability to concentrate, impaired judgment, and slowed reflexes. Even at relatively modest heights of five thousand feet, the reduction in oxygen can cause reduced night vision.

Lack of oxygen is the major problem with high altitude, but there are others: as the air pressure outside him decreases, the ground-pressure air in Nathan's middle ear expands and pushes against the back of his eardrum. He has to equalize the pressure outside with the pressure within his ears to allow air to enter through his eustachian tube. So if Nathan has a bad head-cold, he'll probably want to avoid flying. Ditto if he's feeling gassy.

And then there are the gases dissolved in his blood. It's not a very scientific definition, but decompression sickness occurs, basically, when your blood (and other fluids) begins to boil. Imagine a plastic bottle of soda—before you twist open the lid, it is tranquil in its pressurized state. But as soon as you open it, the pressure decreases rapidly and it fizzes up, with bubbles appearing in the liquid and rising quickly. Now: imagine that your body is that bottle of soda. Gases that are dissolved in your blood come out of solution: nitrogen bubbles cause the bends (which is also a hazard to scuba divers who ascend from the depths too quickly), as well as "the creeps" and "the chokes"—ailments with different locations in the body, and different sensations, but the same cause. If it gets bad enough, or occurs in your heart or brain, it can kill.

Temperature Drop

In addition to getting enough air to breathe, Nathan also has to stay warm. The air gets colder as you get higher. As Nathan rises a football-field high, the temperature goes down by more than a degree Fahrenheit. (Really, we ought to be using the metric system. In that case, I'd tell you that he loses one degree Celsius with every 150 meters in height.) That's barely noticeable if he starts out on a pleasant sixty-degree Fahrenheit day on the ground at sea level. But if he rises farther, the temperature becomes a problem. By the time he reaches ten thousand feet, the temperature would be down to twenty-four degrees Fahrenheit, well below freezing.

Twenty-four degrees is cold, but a New Yorker's winter wardrobe can handle that temperature. Unfortunately, unless Nathan is floating along with the wind, he also has to deal with windchill. On the ground, we measure windchill in how fast the wind is blowing past us, but in the air it depends on how fast Nathan is moving in a direction contrary to the air around him. The faster Nathan flies through the air, and the colder the air is, the more body heat he loses.

So if Nathan is floating along at ten thousand feet, he'll want his winter coat. But if he's zooming around, he'll need more arctic gear. Either way—when Hiro, sitting in the diner, first sees Nathan land in the desert clad only in his boxers, it's amazing that Nathan isn't covered in frost!

Acceleration

And speaking of that scene, it's also amazing that Nathan didn't hurt his bare feet terribly, scraping the skin and burning them on the sand due to friction, landing the way he did. His invisible rockets (or whatever he's using to provide thrust) must have been helping to slow him down as he came back to ground level. Nathan is looking more and more like the Man of Steel.

There's also the question of the acceleration that the normal human body can withstand. Going at supersonic speeds is not necessarily dangerous to the human body, from the point of view of acceleration. (It is dangerous in that you have to manage to breathe and avoid both windchill and windburn.) It all depends on how fast you get up to speed. For a ground-bound example: my little car takes me from zero to sixty in a good twenty seconds, and I barely feel the acceleration. But when a friend with a souped-up sports car takes me out for a spin and goes from stopped to sixty in four seconds, I'm pressed back in my seat (and holding on for dear life). It's not how fast you go—it's how fast you change speed!

So to sum up, what do we have? We have Nathan, a guy with the ability to fly, who—if he's flying much higher than the buildings in his native New York—had better have additional abilities to handle rapid changes in air pressure and in temperature and in acceleration. He'd also want a very efficient metabolism and a good sense of altitude. It also probably wouldn't hurt for him to carry a pocket atlas with him.

And where does that leave us, the more or less normal humans? Are we doomed to fly only from airports, only inside metal contraptions powered by fossil fuels? No, not necessarily. Several flying machines that run on purely human-powered flight have been demonstrated, the first of which was the *Gossamer Condor.*

HUMAN-POWERED FLIGHT

In 1977, pilot Bryan Allen climbed into a small cabin underneath a hundred-foot-long contraption made mostly from thin aluminum tubes covered with Mylar. He controlled the aircraft with his hands and pedaled with his feet. The pedaling mechanism—much like a recumbent bike—provided all the power necessary to get the craft into the air.

The entire *Gossamer Condor,* as the craft was called, weighed less than Allen, at only seventy pounds. Allen took off, cleared a ten-foot hurdle and maneuvered the Condor around two pylons in a figure-eight pattern, cleared a ten-foot hurdle at the end, and landed. It took him seven and a half minutes to go a mile and a quarter—roughly ten miles per hour—but in that time he'd demonstrated human-powered, controlled flight, made possible by both the strong and lightweight materials and the ingenuity of its designers, Drs. Paul B. MacCready and Peter B. S. Lissaman.

The impetus for designing the *Condor* was money: a $50,000 prize offered for the first human-powered aircraft that could cover the course. A few years later, the same designers developed a subsequent aircraft, the *Gossamer Albatross,* which won another prize by flying across the English Channel on only human-supplied power.

The *Gossamer Condor,* which made the first sustained, controlled flight by a heavier-than-air craft powered solely by its pilot's muscles, now hangs in the Air and Space Museum.

The *Gossamer Condor* was a materials-science triumph based on a material (Mylar) invented two decades earlier during a period of great innovation in plastics. The challenge of the prize was also nearly two decades old by the time Mac-Cready and company came to collect.

The same team, more recently, developed unmanned craft that are powered by the sun, with energy collected by solar panels on the wings. Today, polymer technology is still advancing, but so are other areas of materials science. Materials scientists are investigating very strong, lightweight, composite materials as well as nanotechnology and fuel cells. Aerogels, one new material, are very light solids, like a foamy substance but still solid. Depending on what gases are caught inside them, they could be lighter than air. Aerogels are beginning to find use as thermal insulation in places where the weight must be very low or light must be able to pass through the material. They are sometimes called "solid smoke."

Meanwhile, engineers from MIT are developing the gridlocked commuter's daydream: vehicles that convert from cars into airplanes and back. As long as basic technology continues advancing, and as long as people want to fly, our chances of someday soaring more or less the way Nathan does look good.

That takes care of us but what about Nathan? Obviously, he's light on his feet—on the dance floor and off of it. He needs some sort of adaptation to fly, and probably a combination of them. I don't have any clue how he manages all these things without appearing any different from the rest of us (except for his pleasantly chiseled jaw). Maybe his weight is reduced because he has hollow bones, blood that is unusually full of gases, and some sort of organ that can quickly fill and empty a cavity full of buoyant gas? Maybe, also, his takeoffs are assisted by thigh muscles that would make Batman weep with envy, and either his digestive track or some other organ provides him with rocketlike thrust? While we're redesigning his anatomy, perhaps Nathan has adaptations that help keep him warm at high temperatures? This could be sort of like penguins' feet, which manage to keep from freezing even through the Antarctic winter. All this, and he looks good in a suit, too!

Further reading

The Flight of the Gossamer Condor, documentary directed by Ben Shedd, 1978. Smithsonian information on the *Gossamer Condor:* http://www.nasm.si.edu/research/aero/aircraft/maccread_condor.htm.

General Electric's Web site has some lovely flash demonstrations on how jet engines work: http://www.geae.com/education/engines101/.

James Kakalios, *The Physics of Superheroes,* Gotham, 2006. Professor Kakalios teaches basic physics with examples of comic-book superheroes. A wonderful book.

Chapter 5

CLAIRE

Save the cheerleader, save the world. But of all possible cheer-leaders, in all possible worlds, one would think that Claire Bennet needs saving the least. Her ability to heal makes her...resilient. Very.

It's hard to keep a good cheerleader down—and for Claire, these words apply in several senses. Shall we count the ways? She climbs several stories, then jumps off an aban-doned gravel plant. Her broken bones and crushed flesh heal fast enough that she can stand up within seconds. For that matter, Claire seems to intentionally jump from heights fairly often. In the Season One finale, she jumps from yet another height to escape Angela and Peter Petrelli.

Claire also walks into a fire to save a man caught in a train wreck and is both strong enough to rescue him and resilient enough to heal—again, within seconds—from the burns. When Brody tackles her by accident and breaks her neck (an accident that could easily cause paralysis or death), she heals before he notices. Claire also walks away from traffic accidents and an encounter with our favorite serial killer, Sylar, in the high school locker room.

What is healing?

On a cellular level, healing includes removing damaged cells and replacing them with healthy new ones. It can also involve replacing the structural supports for cells. We replace and repair all the time, maybe not on the scale or with the speed that Claire does, but we are constantly growing new hair, new nails, new skin. Less obvious replacement goes on internally, where we can't see it.

In some ways, Claire is much like Marvel's Wolverine—minus the adamantium claws, which is fortunate for her, since they would get in the way of cheerleading moves—but her bones are not unbreakable so much as they are exceedingly quick to heal. Her recuperative powers do seem stymied by physical displacement: she sometimes needs to nudge bones back into something like the correct place.

The other people on *Heroes* with the same ability to heal

from damage that would be crippling or deadly for anyone else—Peter Petrelli and Takezo Kensei (aka Adam Monroe)—have the same limitation: they seem unable to "speedheal" around objects that pierce them, although once the objects are removed, they go back to turbospeed healing. For those of us with normal non-superpower-enhanced healing, if we survive the original injury, we can sometimes accommodate foreign objects. For example, I still have some fragments of glass under my skin from encountering a glass door as a child.

The rate at which we replace cells depends on what type of cells they are, and there are a lot of choices: two hundred to three hundred different cell types are traditionally recognized in the human body. We make new blood all the time. The linings of our guts last barely a week. (Would that the plumbing in my house renewed itself the way my personal plumbing does.) Speeding up our metabolism speeds up growth. So far as I know, the fastest growth is caused by some cancerous tumors, which can grow so fast that their blood supply can't keep up. This leads to areas of dead tissue at their centers.

But that's not to say that we're a mass of uncontrolled cellular growth. Growth is normally carefully controlled. And some cells, like those in our brains and most of our internal organs, are not normally replaced.

Division is one way that we replace cells. On a cellular level, this is cloning. Both the resulting cells contain the same

DNA and do the same job. But another method of creating new cells uses stem cells: unusual cells that can differentiate into a variety of cell types. Stem cells in our bone marrow make our blood cells, for example. Given different signals, these same stem cells can also make cells for a variety of internal organs and tissues.

But while some of our organs renew themselves, we can't heal as much as Claire is able to. An injury to our spinal cords can paralyze us for life. Damage to cells in our pancreas causes Type 1 diabetes, which requires lifelong glucose monitoring and insulin injections. If we step on a land mine, then very likely we'll live the rest of our lives without a foot. Damaged hearts and brains and nerves can recover somewhat—but they don't regenerate. And, on a much smaller scale: get a tattoo and you'll live with the skin modification for the rest of your life.

And in this, mammals are unusual. In Claire's biology class, the students watch a time-lapse movie of a newt regrowing a limb. Newts and starfish grow new limbs, as can tadpoles and embryonic fish and chickens. A wounded zebra fish can regrow almost anything: skin, bone, joints, nerves, arteries, veins, muscle, eyes, spinal cord, and heart. And a spineless little creature called a sea squirt can regenerate its entire body from just a few cells. So, in the greater realm that includes all the animals, Claire's ability to regrow body parts isn't unusual. (She's still a speed demon, though.) She just has an unusual ability for a fairly mature mammal.

Miracle regrowth

When she snips off her pinkie toe in "Lizards," Claire has just had a lesson on regeneration in newts, so maybe she's studying the regeneration process. She's not doing anything brand-new to her: she regrew bits of her hand after she reached into a running garbage disposal back in Texas. And, if she was a first-trimester fetus, she'd also be able to regrow fingers and toes.

But the most limb regeneration that we've documented post-birth is that children can often regrow their fingertips. Internally, we're slightly better off: although our livers seldom replace cells normally, if the liver has been damaged or reduced, our liver cells will divide and regrow to normal size. And while regrowth of arms and legs would be excellent, more people could use regrowable internal organs.

New organs for old

Compared to most of the other characters on *Heroes*, Claire would seem to be in greater danger from the Company or any other unethical group willing to kidnap and use evolved humans simply because her power to regenerate is so desirable, so much needed in the world today. The chance to study her ability in order to reproduce it—or just use it— would be well nigh irresistible.

Why is that? Consider: in the first half of 2007, 16,758 organ transplants were performed. And still, as I type this, there are 97,948 people in the United States waiting and hoping for an organ transplant. That's the same number of people as live in the city of Charleston, South Carolina. And that number is likely to increase as the baby-boom generation ages.

Discussing organ and tissue donations can get a little—or a lot—ghastly. Our bodies are precious and irreplaceable to us, so talking about trading bits of ourselves conjures thoughts of Frankenstein and countless slasher stories. And let's not even mention zombie movies. (I've never understood the zombie taste for brains. Even on a steady brain diet, zombies never seem to get any smarter.)

But look at organ donation from the other side: a living person with one failing organ could regain their health if a replacement could be ethically obtained from a person or animal. And we've learned a lot about how to do this: the first liver transplant occurred in 1963, and the first heart transplant in 1967. Today, most organs are from people: specifically, from recently deceased bodies of people who, during life, decided to donate. Despite my sentimental and literally visceral attachment to my body while I am alive, I don't see any reason why organs from my corpse shouldn't be used to save other people's lives. When the donor is dead, most of the risk is on the recipient's end. Will her body reject the new tissue (since her immune system very properly recognizes that it is made of alien cells)? Will the drugs she takes for the rest of her life, in order to suppress her immune system, cause other medical problems? With her immune system suppressed, will she die of a different infection?

Some organ and tissue donations can come from living people. These donors are taking additional risks with their health, but they can weigh the risks and give informed consent at the time of donation. Live donors can contribute one kidney, part of the liver (since as mentioned above, healthy livers can regenerate), skin (which also regenerates), part of the pancreas, part of the intestines, bone marrow, and part or all of a lung.

Oddly enough, a living person can donate a heart in some circumstances. Sometimes, someone with damaged lungs but an excellent heart has a better chance of surviving a lung transplant if he receives a heart-and-lung combination. In that case, the healthy person donates a heart and a lung. The ill person's heart (and unhealthy lung) is removed, but the heart is installed in the healthy person. They swap hearts, in a sort of surgical Valentine's procedure.

For most organs and living donors, the donor can only make a single donation. But Claire—if she consented—would be an über-donor, easily able to regrow both solid organs and tissue. But even with her healing abilities, one woman couldn't ethically be asked to act as a donor to anything like all the people in the United States who need transplants: if she donated a hundred times a day, every day of the year, she'd still spend about three years donating to just the people who need a transplant now. On the other hand, Claire could only donate to folks who had the same six immunological factors that she did, since no one is a universal donor for organs, unlike blood. And without a Claire, we are in an even more dire situation.

There must be a better way. As the situation stands now, many potential recipients die while on the waiting list. Organ donation groups are trying to increase participation. An entire field of research is investigating ways to grow organs specifically for donation. And regenerative medicine is a new, and very cool, field that may allow bodies to heal themselves, without needing extra bits from elsewhere.

We don't have a superhero willing to grow us organs, or enough human donors to keep up with demand. Physicians are looking for other options. Xenotransplantation is one possi-

RISKY BUSINESS

In the episode "Four Months Ago," Peter learns that Adam's blood can heal the horribly burned Nathan. Saving his brother becomes Peter's impetus to escape the Company, and take Adam with him.

In Nathan's hospital room, Adam injects blood (presumably his own, newly drawn blood) into Nathan's IV drip. Within a scant second or two, Nathan's burns heal.

This is problematic for a couple of reasons. Just as with organs, there is a chance that Nathan's immune system will violently reject the blood—so instead of healing Nathan, it could kill him. No physician in the civilized world would allow a blood transfusion without first testing for a match and type.

Second, why does Peter need Adam at all? Once he recovers his powers, Peter has the same ability to heal that Adam does. And, since they're brothers, Peter's blood is more likely to be compatible with Nathan than Adam's.

But okay, it is possible that Peter believes that Adam has a different power, and they must use Adam's blood. Fine. Only...is it? Assuming that Adam's ability is the same as Claire's—ridiculously fast regeneration of cells and tissue and, effectively, immortality for the organism as a whole—then what's likely to happen inside Nathan? His system, which is already under a lot of stress, suddenly sees alien proteins and cells from Adam's blood. Nathan's immune system attacks the invaders, as we might expect.

But Adam's blood is capable of fighting back. Adam's white blood cells are initially far outnumbered, but they (like the rest of Adam) can reproduce incredibly quickly. So then Adam's immune cells are likely to fight anything they perceive as alien: that is, all of Nathan. Not only that, but the same mechanism that allows the cells on the site of Claire's snipped pinkie to regrow bone and nerves and muscles and skin and even a toenail may allow Adam's blood cells to grow all sorts of other types of cells. (We'd call those totipotent stem cells—I'll get into that in a bit.) And do so *incredibly quickly*.

This is all speculation, of course. Real healing doesn't work this way. But, given what we're shown about this superpower...wouldn't injecting Adam's blood into someone turn that person into a clone of Adam? Or, maybe if it doesn't...has the injection given Nathan the power to heal? If so...maybe the shots fired at Nathan at the end of the episode "Powerless" aren't quite as deadly as they look?

bility: using organs from different animals in humans. Organ transplants from other primates have occasionally succeeded, while organ transplants from mammals less related to humans have not. A lot of doctors think that pigs may be the answer.

Tissue from pigs has been used for several decades to repair human heart valves. Pigs are also the source of some insulin used by diabetics. A number of groups are researching cell therapy, injecting cells from pigs (or other animals) into humans to treat diseases. For example, instead of injecting insulin that was originally created in a pig's pancreas, why not inject some cells from the pig's pancreas and let the cells create insulin inside the diabetic's body? (This only works with Type 1 diabetes, by the way.) As long as the cells stay alive and produce the correct hormones, the diabetic could be freed from daily injections of insulin.

The main problem with transplanting pig cells is that the patient's body rejects the implanted organs, sometimes within minutes, sometimes more slowly. Then the challenge becomes finding a way to keep the pig tissue alive while also keeping the patient healthy. Some researchers are trying to create pigs that carry a human protein to make immediate rejection less likely. Other researchers are experimenting with different methods of encapsulating cells so that they can receive nutrients and produce insulin (for example) while being hidden from the antibodies that would usually attack them. A company called Living Cell Technologies may (given ethics approval) start human trials on a version of this work this year.

FARMING ORGANS

Sheep are not, broadly speaking, all that different from humans, in that they are fellow mammals. Professor Esmail Zanjani at the University of Nevada is investigating whether sheep can be used to grow human organs to replace damaged and failing ones. If so, the sacrifice of a sheep could save human lives. Given that many of us eat meat from livestock, the ethics of sacrificing a domesticated animal to save human lives is largely settled.

In biology experiments, test animals are often injected with human cells, but they don't tend to grow all that well, since they are usually seen as invaders and trigger the animal's immune system, which sets out to find and kill alien cells. But Zanjani and his coworkers discovered that they could inject a particular type of human cells (hematopoietic progenitor cells) into "immunologically naive" sheep fetuses. These cells were accepted, grew, and differentiated into a variety of different types of cells. The final result was adult sheep in which about 15 percent of the animal's cells had human DNA.

That's not quite the final goal of growing all-human organs inside livestock. But even individual cells might be helpful for cell therapy. The human cells could be extracted from the sheep cells—commercially available cell-counting machines could be modified to do that—and used for humans that need them.

FURTHER READING

G. Almeida-Porada, et al, "Plasticity of Human Stem Cells in the Fetal Sheep Model of Human Stem Cell Transplantation," *International Journal of Hematology* 79 (January 2004), 1–6.

And then, there are the groups that would rather grow entirely human tissue in animals.

Part-human, part-sheep chimeras strike many people as creepy—as creepy as Frankenstein's monster or urban legends of Bat Boy. But for some reason, fewer people are bothered by transgenic mice, which are an enormously helpful tool for biochemists.

Meanwhile, other people believe that it is unethical to use animals at all for medical research. And even if xenotransplantation becomes a successful method of developing new organs, there are still risks that a disease in the animal might cross over into humans. Avian flu is one example of a cross-species disease, but it could be even more dire in patients with suppressed immune systems. If we could, instead, grow replacement parts in test tubes, we would avoid the ethical question of using another creature for our benefit. Even better would be medicine that could convince our bodies to regrow and correctly place cells that we need.

Tissue engineering

In some cases, we can already grow tissues in the lab. We can grow skin by first taking a small sample, then encouraging the skin to grow in a lab. This larger piece of skin can then be grafted back onto a person—who may have been burned, for example.

GROW YOUR OWN BLADDERS

Three boys and four girls are the first people in the world to receive laboratory-grown organs. The children, aged four to nineteen, received bladders grown from their own cells in 1999 through 2001, and a four-year follow-up shows that the organs are doing fine.

Bladders are relatively simple organs: they hold pee generated by the kidneys, and have three layers: an inner layer of urothelial cells that line the entire urinary system, a layer of collagen, and an outer layer of muscle cells. But when they don't work right, it's both a physical and social misery because the person can't control urination and the condition can cause potentially life-threatening damage to the kidneys.

Back in 1999 through 2001, children whose bladders had not developed normally were part of an experiment at Boston Children's Hospital. Usually, children with this condition would have surgery, and a piece of their intestine would be extracted, and shaped into a bladder. One problem with this procedure, however, is that intestine tissue is designed to absorb nutrients, which is not the best characteristic for a bladder, and can cause more problems over time.

Instead, surgeons removed a small sample of each child's bladder, teased apart the layers, and started growing the inner and outer layers, separately, in the lab. After they had enough cells, the layers were placed onto the appropriate side of a biodegradable mold shaped like a bladder. About two months after the first surgery, the lab-grown bladders were sewn onto the patients' original bladders.

The new bladders worked as well as intestine-based bladders, without the associated side effects, and they have continued working.

FURTHER READING
A. Atala, S. Bauer, S. Soker, J. Yoo, A. Retik, "Tissue-Engineered Autologous Bladders for Patients Needing Cystoplasty," *The Lancet* 367 (2006), 1241–1246.

Stem cells

I mentioned earlier that some cells divide into like cells. But special cells can produce multiple types of cells. These are stem cells, so called because, like the stem of a plant that supports leaves and flowers and fruit, these cells can produce many different, specialized types of cells.

The existence of stem cells is inevitable in some ways: any species that replicates by fusing an egg and a sperm into a single cell (fertilization) had better show serious versatility in going from that first—and highly specialized, in its own way—cell to the hundreds of different types of cells necessary in a newborn. In embryos, this differentiation is fueled by a specific type of parent cell: an embryonic stem cell.

You've probably heard of these cells. There's a lot of disagreement over the ethics of using them. Human embryonic stem cells are derived from human embryos, five days after the egg and sperm cells fused. If the fertilization had occurred in a woman's body, the embryo would not yet be ready to implant in the wall of her uterus. But the embryos that are used are fertilized in a lab dish.

What was a human embryo doing in a lab dish? It's there because of a technique called in vitro fertilization (IVF), which is used to help couples conceive a child. From a cellular point of view, the usual method of fertilization is that the woman's egg cell and the man's sperm meet and fuse in the woman's body, between her ovaries and uterus. The fertilized cell starts to divide into more cells as it travels down to the uterus. After several days, the embryo implants into the lining of the uterus and continues to grow into a fetus. But at each step along the way, the process can be derailed: a ripe egg and sperm may not meet; even if the sperm and egg meet, they may not fuse; the embryo may not grow normally; and/or the embryo may not implant into the wall of the uterus.

When you look at it that way, it's surprising that anyone is successfully conceived at all. (It also makes the population explosion even less explicable.) Then again, an average woman releases about four hundred eggs in her lifetime—far more than she could possibly carry to term. Meanwhile the average man has an even greater overabundance: a typical man produces about four hundred billion sperm in his lifetime.

Nevertheless, sometimes in vivo (in the living body) fertilization runs into trouble. In vitro (in glass) fertilization can help an infertile couple to have a child by taking part of the fertilization process out of the woman's body and into the lab. In 1978, Louise Brown was the first such "test tube" baby (actually, a "petri dish" baby, but that doesn't roll off the tongue as easily). Here's how it works:

First, the woman takes drugs that stimulate her ovaries to ripen many egg cells at the same time, rather than the usual one per month. Then she has surgery during which a number of eggs, up to about ten, are removed from her ovaries. Each egg is put in a lab dish with the man's sperm. If all goes well, the sperm fertilizes the eggs within two or three days.

Sometimes these early-stage embryos develop abnormally. Those embryos are discarded. Of the normal-looking ones, one or two or sometimes even three embryos are reintroduced to a woman's (not necessarily the same woman that donated the eggs) uterus. Each embryo has about a 20 percent chance of embedding. At the point that they are reintroduced, the embryos usually have between two and eight cells.

This leaves as many as eight or nine embryos sitting in the lab dishes. We have the technology to keep them alive in the lab for a while, by freezing them, but not to grow them larger. Since the drug-and-surgery combination can be painful, the IVF is costly, and the chance of failing to obtain a viable pregnancy is fairly high, couples tend to keep the frozen embryos, which can be thawed to provide second and third chances, rather than starting the procedure all over again. (Sperm cells can also be frozen, but thus far, ripe egg cells cannot.)

The ethics of in vitro fertilization are one area of argument today. The ethics of how embryos fertilized in a lab dish should be treated are another. There is no question that many naturally conceived embryos die without human intervention. But once humans do intervene, we face many more questions! Under the right conditions, this collection of eight cells can become a human: so what responsibilities do we have, if any, toward that embryo? The embryo contains human DNA, so should it be treated with at least the same respect as adult human body parts? Who has legal and ethical rights and responsibilities toward it: just the egg and sperm donors? How about the person and organization who oversee the fertilization? Can it be owned?

I don't know the answers to those questions, and our society hasn't reached a consensus yet either. What we do know is that human embryonic stem cells are obtained from five-day-old, unimplanted human embryos that were fertilized in a lab dish. These embryos cannot continue to develop there: without further action, they all die quickly. Even if the most proactive efforts are taken and the embryos are introduced to a woman's uterus, 80 percent of them will still die.

Given the ethical questions it raises, why would anyone want to do anything with these embryos? For two reasons. First, studying embryonic stem cells can teach us how embryos develop, which could lead to a better understanding of the diseases based on abnormalities, including birth defects and cancer. Second, embryonic stem cells could perhaps be used for therapy: they could perhaps be used to generate tissue for organs that don't usually regenerate, including brains and hearts. Potentially, embryonic stem cells could be used to treat Parkinson's and Alzheimer's diseases, strokes, spinal-cord injuries, burns, heart disease, diabetes, and some types of arthritis.

They are attractive because embryonic stem cells can differentiate into all the types of cells in the body, and because they can be grown in the lab more easily than adult stem cells. Adult stem cells are found within differentiated tissue, and they can produce only limited types of cells. For example, stem cells in our bone marrow produce a variety of blood cells. So far as we know, they can't change their characteristics.

One intriguing possibility is that perhaps adult stem cells can change to act more like embryonic stem cells. When newts and salamanders are hurt, their adult cells can change back into something like their embryonic state, and then divide to regrow the lost body part, whether it's a leg or tail—or nerves. If we could learn how to make human cells do that, then adult stem cells could be used to regrow severed spinal cords, undo other nerve damage, and perhaps brain damage from injuries and disease as well.

And there isn't a hard-and-fast line between embryonic and adult stem cells. Investigators at Wake Forest University and Harvard Medical School also recently discovered that amniotic fluid—the watery pouch that surrounds a growing fetus—contains stem cells that show characteristics of both. The fluid they examined was left over after amniocentisis—a medical test that examines the fluid to determine the health of the fetus. These cells were coaxed to turn into muscle, bone, fat, blood vessel, nerves, and liver cells.

We do have another problem, though: nobody has figured out how to get human embryo cells to grow without using mouse feeder-cell layers to stimulate growth. So human stem cells grown in vitro are contaminated with DNA from mice.

For Claire to be able to heal and regenerate, her cells must show some sort of stem-cell-like properties. Plus, her cells must use up extraordinary amounts of energy, in order to have the energy to divide and grow as fast as they do.

Healing from the outside

Daniel Linderman also shows an ability to heal—but he can apply it to other people and even plants. Merely by holding Heidi Petrelli's hand for a moment, he's able to heal her enough that she can stand again. In some way, he causes others to do what Claire does naturally. Maybe Linderman would be a better answer to the organ donor problem than Claire? One wonders whether he may have tried such healing, and was so overwhelmed by the resulting demand that he was forced to consider whether having fewer humans would be better than saving humans?

In the online graphic novel *War Buddies*, which highlights the background of Linderman and Arthur Petrelli, they encounter a girl called Au Co, who can cause entire fields of plants to grow extraordinarily rapidly. Growth and healing are intimately connected—two sides of the same thing.

✳ DNA damage

In "Company Man," Claire survives both being shot by Matt Parkman and radiation damage from Ted Sprague. These are very different types of damage. A bullet causes physical damage (and a lot of it). But radiation damages the DNA—the instructions in each cell that tell it how to act.

In normal people, damage to DNA can cause cells to die, or just become less efficient, or to reproduce incorrectly— sometimes leading to cancer. But our bodies also have a number of ways to repair damage to DNA. So maybe Claire's ability to recover from DNA damage is just a turbocharged version of a normal ability. Claire's ability to regenerate may also prevent her from experiencing some of the damage that causes aging.

Memories

The Haitian doesn't take Claire's memories when he is ordered to, in the episode "Fallout." And although he succeeds in taking Peter's memories in "Four Months Ago," Peter's ability to heal allows him to regain the memories.

Claire's ability to heal hasn't stopped her from growing and maturing, and it doesn't stop her from creating memories. That's impressive healing. But it goes further: in "One Giant Leap," Claire fights off Brody's attempt to rape her, but she falls and a branch impales her brain. She appears dead,

not just to the iniquitous Brody, but also to the medical work-
ers who find her body. Certainly, no pathologist would start
doing an autopsy unless she was very sure that all the functions of life have stopped. Claire also appears to die when
Matt Parkman shoots her, in "Company Man." And yet in
both cases, she's not as dead as she seems.

She's *not* dead, Jim

Dead is different from "very badly hurt." When a person
dies, not all the cells necessarily die at once (if that were so,
organ transplants wouldn't be possible). But even though bits

---

WHAT, PHYSICALLY, ARE MEMORIES?

When you make memories, what's going on in your brain? Unlike
your computer's hard drive, a brain doesn't have a section dedicated
to storing memories—they seem to be spread across the brain. Moreover, there are different kinds of memories. And even a single memory
isn't localized: it appears to be a pattern of connections between neurons in different parts of the brain. Creating a memory appears to
change both the physical structure of the connections between neurons and the chemicals produced by parts of the brain.

FURTHER READING
Howard Eichenbaum, *The Cognitive Neuroscience of Memory: An Introduction*, Oxford University Press, 2002.

---

may be healthy, the body as a whole is incapable of the organized functioning necessary for continued survival. A doctor, generalizing broadly, explained it as, "They're dead when they ain't coming back."

By the time Claire was in the morgue, the intricate mechanical and chemical processes that power the individual cells of her body should have broken down. Her heart had ceased to pump, her lungs ceased breathing, her guts ingesting; her circulatory system was no longer carrying oxygen and nutrients to individual cells and carrying waste products to her kidneys. The delicate ballet of electrical and chemical signals that pass through the blood and neurons had fallen into disarray as the tissues and organs starve.

Life, the alien Time Lord of *Doctor Who* once said, is just nature's way of keeping meat fresh. By the time the branch is removed from Claire's brain, her tissues surely need more than just refreshening: they need rebuilding.

At least Claire had time to herself to recover—if the pathologist had gone beyond the initial incision to removing the internal organs, Claire would have had to grow entire organs from scratch. And if she could survive that, then perhaps she could survive Sylar cutting open her head and doing whatever it is that he does with the brains of his victims. The workings of the brain are mysterious—but so are the workings of our livers. It may be mysterious, but it isn't magic.

This is all very reminiscent of the fantasy movies and TV show *Highlander,* in which the "immortals" cannot die except by decapitation. In *Highlander,* there is apparently something magical about the spinal cord at the neck. In *Heroes,* there is apparently something magical about the brain, all indicated by Claire's temporary death-by-wood-in-the-brain, Peter's temporary death-by-glass-in-the-brain, and Sylar's fixation on the brains of his victims.

On the other hand, let's speculate: suppose that Claire's cells, in addition to being super energetic, are able to go into some sort of hibernating state. The damage from the branch may have disrupted parts of the brain stem that are responsible for regulating a lot of the body's functioning, including maintaining blood pressure. Perhaps the branch physically disrupted her autonomic nervous system (the part of the nervous system that regulates involuntary action, like in our guts and glands). If removal of the branch is enough to rouse the cells in her brain, and they have somehow stored enough energy to heal the damage, then restoring her autonomic system would start sending signals out to her heart to pump, her blood vessels to dilate or contract to send blood to her brain, and increase her body temperature... This gives an entirely new meaning to the platitude about "she's not dead, she is only resting."

Just as Kensei has aged well—or rather, apparently not aged at all—perhaps Claire will stay looking sixteen forever.

And she'll have the company of her uncle Peter, who now shares her ability to heal.

A little more oomph

If she wanted to play extreme sports or be a crime-fighting superhero, Claire might value her abilities more. But as a teenager trying to find her place in society, she would much rather be normal. And her healing abilities are normal—most of her ability is just normal physiological functions with a little extra oomph.

As we learn more about the processes that go on in our bodies, and develop new biochemical and biomechanical tools, science may provide that extra oomph to us as well. At best, Claire may not be abnormal—she may be just leading the way into a healthier future for us all.

Further reading

B. Alberts, et al, *Molecular Biology of the Cell*, 4th edition, Garland, 1994.

M. D. Dooldeniya and A. N. Warrens, "Xenotransplantation: Where Are We Today?," *Journal of the Royal Society of Medicine* 96 (2003), 111–117.

Regenerative Medicine 2006, by multiple authors, is published online by the National Institutes of Health. Available (with updates) at http://stemcells.nih.gov/info/scireport/2006report.htm. More information about stem cells is also available from NIH at http://stemcells.nih.gov.

18 Ways to Make a Baby, NOVA program, originally broadcast on October 9, 2001. Associated information on the Web at http://www.pbs.org/wgbh/nova/baby/.

Chapter 6

CLAUDE
"CAN YOU SEE ME NOW?"

It's awfully hard to fight someone you can't see. Claude—*Heroes*' invisible man—uses this to his benefit, whether he's shoplifting, spying, or escaping a bespectacled, gun-toting Mr. Bennet. And even though he disappears every time Peter Petrelli wants to find him, he keeps turning up when Peter really needs him. Grouchy, disillusioned Claude can turn invisible, but his motivations are far from transparent.

Claude is, of course, not the first character to be invisible: H. G. Wells wrote about a mad scientist turning himself invisible in 1897, and Claude's namesake was Claude Rains, who played the invisible man in the 1933 movie. That character, Griffin, is a chemist who creates a potion that turns his entire body utterly transparent: his refractive index changes to be the same as air, and doesn't absorb or reflect visible light. But once he makes himself invisible, he can't change back. Perhaps because he's sniffed too many mercuric fumes or because he realizes that his invisibility (not to mention criminal activity) has cut him off from normal human society, he goes mad. Only after he's bludgeoned to death does his body become visible again.

Wells's story unleashed a spate of invisible men, women, spies, and yes, even an invisible dog, with basically the same explanation. From *The Fantastic Four*'s Sue Storm to Harry Potter's cloak, the notion of invisibility has sparked a multitude of possibilities. But how realistic is any of it?

Chameleons, cuttlefish, and Claude

Claude can become undetectable to the human eye. But how could that be possible? There are many illusions used to trick our eyes into showing us things that don't exist and to hide things that *do* exist. It isn't all done with smoke and mirrors—but they can help. In the natural world, plenty of animal predators and their prey have evolved some form of camouflage. And a few animals actively change their appearance to fit the situation. Chameleons are famous for changing their skin color to match their surroundings. Some ocean animals are even better at blending in. Claude disappears in less than a second, which is too fast for a chameleon. But that speed is no problem for a cuttlefish.

Cuttlefish, which are related to squids, are quick-change artists. With a complex multilayer skin and as many as two million multicolored pigmented cells in their skins—as many as two hundred per square millimeter—they can replicate any color of the rainbow within moments. Not only can they copy the colors around them: they can copy patterns from their surroundings.

Cuttlefish watch their surroundings and adapt to fit in. To help do that, they have a bizarre eye anatomy. They can see very well, which helps, of course. Their eyes have W-shaped pupils and eyeballs that change shape to change focus. They can also look both in front of and behind themselves—at the same time, with the same eyeball! We're used to eyeballs that have one retina—one place where light is focused at the back of the eyeball, but cuttlefish have two high-sensitivity spots of photoreceptors. That's pretty hot stuff for an animal that never evolved a backbone!

Maybe Claude doesn't turn invisible so much as use very effective camouflage? If he had the skin of a cuttlefish, all he would need to do to nearly disappear is make his front look like whatever is behind his back. Of course, if there is more than one person around, he'd have to keep up appearances from a couple of different angles. Walking through a crowd could get complicated.

Still, camouflage is a pretty simple idea that is surprisingly effective. At the University of Tokyo, Professor Susumu Tachi took a picture of the wall of a room, then projected that picture onto the wall. Then he made a hooded jacket out of a material like that of a movie screen. When a person wearing the jacket stood in front of the wall, the projected image hid him pretty well. Only his face stood out, and that was probably because we humans are hardwired to find faces. With new flexible displays, the jacket could display the image itself and a projector might not even be necessary.

Claude might not be the only person on the show with cuttlefish skin—Candice Wilmer's shape-changing talent might resemble another cuttlefish ability. Candice can change her features to resemble other people. She imitates Simone, Mrs. Bennet, and Claire, among others. How could she change the shape of her face? In the same way that a cuttlefish changes its shape: muscles under the skin allow it to swell, spike, and change its surface texture from smooth to spiny. To a cuttlefish, an instant nose job is no problem. That explains facial features—and the skin color change would be trivial with cuttlefish-like skin.

But what about Candice's ability to change her hair and clothes? Similarly, how can Claude turn a wallet invisible when he touches it?

Who you can't see can still pick your pocket

Plenty of superheroes have had to deal with clothes and other objects turning invisible with them. In *The Fantastic Four* comic book, Sue Storm (the Invisible Woman) was able to bend light waves, generating a field of energy around her to render herself and her clothes and other things she touched invisible to the naked eye. That sounds like more than usually ridiculous technobabble, but in a very limited way it is not only possible, it has been demonstrated. Bending light rays isn't hard: we've become masters of controlling light, and we're learning more every day. By light, I mean not just the colors that we can see with our unaided eyes, but electromagnetic radiation, which includes X-rays, infrared light, and radio waves as well as visible light. We manipulate light using lenses and capture it with our cameras for applications ranging from microscopy to astronomy.

The newest research is providing us with new ways of bending light. In the microwave part of the electromagnetic spectrum, at wavelengths about a thousand times longer than visible light, a group of scientists showed that they could bend light waves around an object, creating a pocket of invisibility.

THE REAL INVISIBILITY CLOAK

Think about invisibility for a minute. Things that are transparent are invisible, right? Well...no. Not really. Suppose you go out to lunch and a waiter sets a clear glass full of clean water down in front of you. Both the water and the glass are transparent, but if they were invisible, you would have a very thirsty meal since you wouldn't be able to find the glass to drink from it.

But, you say, maybe the glass isn't completely transparent. When we say we "see" the glass, we mean that we see scratches on the glass and smudges left by the waiter's fingers (we are dining at a rather grimy

... continued from pg 145

greasy spoon)—not to mention the straw sticking out the top. And that is true—you can see imperfectly transparent parts of the glass, the same way that it's easier to spot dirty glass doors than clean ones.

That doesn't entirely explain the glass of water, though. What we really see when we look at that clear glass of water is the difference in the "index of refraction" between the air and the glass: when light enters the glass from air, it changes direction slightly. When it exits the glass back to air, it changes direction again. That's why the straw appears to be bent even though we know that it's straight. And whatever we're looking at through the glass appears distorted. Anything that light can go through has an index of refraction that can be used to figure out how much it will bend light. Vacuum, by definition, has a refractive index of one. Air is very nearly one, as well. The clever people who design lenses for binoculars and video cameras and microscopes use different materials to make light bend just the way they want. But of all the materials we've found in nature, every single one has a refractive index above zero. That means that the light always bends in roughly the same direction.

But! There's no rule that says every material must be that way. A Russian physicist, Victor Veselago, figured out that materials with a negative refractive index are theoretically possible, way back in 1968. These materials could be used to create wildly different optics—if we could find them.

We haven't found them—but we're learning how to make them. In 2001, a group at the University of California, San Diego, demonstrated a negative-index material. Now that we know that it is possible, the race is on to make and use these materials. The San Diego group made a "metamaterial" using an optical trick: if you sandwich together materials in many repeating layers much much smaller than the wavelength of light, then the light interacts with the stack as though it were a single material with properties different from either of the component materials.

Because we're not too good at engineering very tiny things yet, the San Diego physicists didn't try working at visible light's short wavelengths (less than one-millionth of a meter). Instead, they worked with radio waves that are closer to three centimeters long. Also, they manipulated the properties of the metamaterial not just by repeating layers but by making regular repeating designs within the layers. They used fiberglass and copper to make the metamaterial with an optical property entirely unlike fiberglass or copper: it had a negative index of refraction.

So, now you have positive and negative index materials and you can bend light any way you want it: you should be able to make a material that channels light so that an area is, literally and rigorously, invisible! Anything could be hidden inside that pocket, and it would cast no shadow, cause no distortion, and it would appear to have a refractive index exactly the same as air. Like water flowing past a stone in a stream, light slips around the pocket. If you could use that sort of material to make your drinking glass, then, when you looked at it, you wouldn't see any distortion. Your glass could even be full of milk, and you still wouldn't be able to see anything!

"It would act like you'd opened up a hole in space," says Professor David R. Smith at Duke University. "All light or other electromagnetic waves are swept around the area, guided by the metamaterial to emerge on the other side as if they had passed through an empty volume of space."

Has anyone made one of these? Not for milk, yet. But in 2006, Smith and other researchers at Duke University and Imperial College, London, made a device that worked like that for radio waves. Right now it's limited to the lab, but they are working on extending it to the real world. Once we're better at building nanotechnology, there is no reason why it can't be extended to visible wavelengths.

FURTHER READING

D. Schurig, et al, "Metamaterial Electromagnetic Cloak at Microwave Frequencies," *Science* 314 (2006), 977–980.

Maybe Claude is literally invisible—then again, maybe he isn't. Another possibility is that he isn't invisible or camouflaged at all: maybe his power is to simply make people overlook his presence: unperceivable rather than invisible. This would also provide a fuller explanation of how Candice's superpower works.

Candice's abilities stretch beyond her own body—she makes illusions of a beach for Sylar in the episode "Kindred" and even hides his perception of the scars on his chest.

To understand the difference between sight and perception, think about what "seeing" involves. We humans use sight as our primary sense, and it is extraordinarily versatile. We can see things just millimeters away from our eyeballs as well as things as far away as the Andromeda galaxy, which is located two and a quarter *million* light-years away.

So how does a person see? Light either radiates from an object (like a lightbulb) or bounces off it (like a shiny apple), travels across space to her head, passes through her helpfully transparent adjustable bio-optical focusing system (that is, eyeballs), and arrives at a light-sensitive retina. There, photosensitive cells absorb the light and send electrical signals to the optic nerve endings. These are routed through a couple parts of the brain that process different types of information from the input and eventually to the part of her brain that assigns meaning to the signals. If any link along that chain of events is severed, her sense of vision fails.

Seeing ain't believing

It is possible to take advantage of the way that we process vision. Assigning meaning to the things that we see, as we see them, is a huge job. Actually, it's an impossible job.

To make a useful model of what we see, our brains have to reverse-engineer reality based on the insufficient information received from our eyes. We can only see in a tiny portion of the electromagnetic spectrum, we can only see from a very limited point of view (we are all stuck inside our own heads), and we receive huge amounts of extraneous information. Given the situation, our brains make reasonable assumptions that serve us very well, most of the time. Most of us can spot a moving car, process its speed and direction, and leap out of the way (if necessary) in a split second. But the process of optimizing vision involves our brain taking shortcuts. To name just two examples: we look for faces and we also track fast-moving objects without a lot of the intermediate steps that we use to puzzle out more complex scenes. These shortcuts to our way of perceiving—as well as our habitual ways of assigning meaning to what we view—can make us think that we see things that don't really exist.

This isn't breaking news: in the 1600s, René Descartes used the fallibility of vision as a reason to mistrust our senses in general (and it was a vital step in setting up the boundaries of the rest of his discussion, including the important one: "I think,

WHY WE DON'T SEE WHAT'S IN FRONT OF OUR NOSES

There are none so blind as those who will not see. We tend to think that we notice everything in our field of vision, but that's not the case. It turns out that humans are surprisingly good at "ignoring the elephant in the room." This is called "sustained inattentional blindness" and it has implications for many things we do, from trying to drive safely while talking on a cell phone to judging the reliability of eyewitnesses in a trial.

In 1999, psychologists Daniel J. Simons and Christopher F. Chabris at Harvard University filmed a one-minute clip of a three-on-three basketball game. They asked the viewers to count how often one team passed the ball. At the end of the clip, most viewers could tell them a number. But only half of the people who watched the film had noticed a gorilla appear at one edge of the screen, walk across the room, thump its chest, and (after spending a total of nine seconds in view) walk offscreen.

Since then, Simons (now a professor at the University of Illinois' Beckman Institute) and others have discovered a whole host of conditions under which we simply are unaware of things that we would expect a normal person to see. For example, in addition to change blindness—which is not noticing something that changes in a scene if you're more concerned with something else that changes at the same time—Simons et al. discovered change blindness blindness: in which individuals don't believe they could be affected by change blindness. Also, Steven Most at Yale and others noticed that the most influential factor in whether a person noticed something was whether they were looking for it. So not only do we sometimes "ignore the elephant"—we are also prone to "ignoring the elephant swinging its trunk" if we're looking for a mouse.

FURTHER READING

The UIUC Beckman Visual Cognition Lab Web site: http://viscog.beckman .uiuc.edu/djs_lab/.

Michael Shermer, "None So Blind: Perceptual-Blindness Experiments Challenge the Validity of Eyewitness Testimony and the Metaphor of Memory as a Video Recording," *Scientific American*, March 2004.

therefore I am" in his *Discourse on the Method*.) And although we can reason that it is all done with smoke and mirrors, there are few things that magic-show audiences love more than optical illusions. Misdirection, expectation, and quirks of visual processing can all be combined to trick us into seeing things that don't exist—or not seeing things that do exist. (I'll talk a little more about illusions in Chapter 11.)

Winston Churchill supposedly said, "Men occasionally stumble over the truth, but most of them pick themselves up and hurry off as if nothing ever happened." And that is more accurate than you might think. We are, it turns out, terrifyingly good at ignoring things that we *don't* expect to see.

So Descartes was right—we can't trust our vision. Especially not when Claude is around. Maybe the radio-age superhero, the Shadow, isn't the *only* one who knows what evil lurks in the hearts of men, maybe he isn't the only noir antihero who has the power to cloud men's minds?

Claude is invisible, whether because he blends in with the background, or wraps light around him like a cloak, or plays a Jedi mind trick to keep people from noticing him. Whichever tactic he uses, it isn't infallible: Mr. Bennet catches up with his old partner using infrared night-vision goggles and a Taser.

Clearly, Claude has been out of sight, but not out of mind.

Chapter 7

D. L. HAWKINS: MATTER IS NO BARRIER

D. L. Hawkins loves his family. He's had a tough life, but he'll do just about anything for Micah and Niki. To take care of them, he'll go through hell and high water. To protect them, he'll go through just about anything. Literally. Prison walls, locked doors, cars, anything.

Unlike most of the other superpowered characters on *Heroes,* D.L. already understands his ability to pass through solid objects when we meet him. And he's shared that knowledge with his family.

To understand why D.L.'s "phasing" through walls is a pretty cool superpower, we need to understand more about matter.

What are the phases of matter?

The matter on Earth comes in one of three forms: gases, liquids, and solids. No matter what kind of atoms and molecules—thallium, thullium, tellurium, or something else—matter is made from, it will be in one of these forms. (There are actually two more types of matter: plasmas and condensates. The latter are very rare on Earth. Plasmas, while seldom occurring on Earth, are inside every fluorescent light and neon sign—and inside our plasma TVs.) One of the easiest examples to describe is water: we encounter this common molecule of H_2O as gaseous water vapor, as liquid water, or as solid ice. Just from our common experience, it's clear that ice is colder than water and water is cooler than, say, steam.

The difference between states of matter depends on how much kinetic energy the atoms or molecules have. And the state of matter dictates how freely the molecules can move and how they are arranged.

Kitchen science

That's a pretty abstract paragraph, so let's put it another way. Suppose you go to the freezer and get one or two ice cubes, put them in a jar, and screw a lid onto the jar. You pretty much know what to expect from ice cubes, just from experience. The solid water in the jar is cold (in other words, it has low kinetic energy). The ice doesn't change volume (that is, the molecules stay about the same distance from one another). The ice also keeps its shape (that is, the molecules stay in the same position to one another). These are the three properties of matter. None of this would surprise an average preschooler.

But then some of the energy from the warm air and warm jar transfers into the ice: the temperature of the air and jar decreases some, and the temperature of the ice increases some, and eventually the ice melts. Now you have liquid water, which has more energy than the ice had. Unlike the ice, the liquid does change shape—the molecules are not as tightly bound to one another as in the solid, and it conforms to the shape of the lowest part of the container. Like ice, the liquid water doesn't change volume (or, at least, not much).

Back to our jar of water. Say that it's a hot day and you put the jar in the sun. The water absorbs more energy from its surroundings and evaporates. You can't see it, but the water is still in the jar—only now it is a gas. Actually, you probably can still see some liquid water, either on the bottom of the jar or as drops condensing on the sides and lid of the jar. That has to do with the saturation point of the air that was in the jar since the beginning. Suppose that when you sealed the ice cubes into the jar, you also sucked out the air with a pump (like the ones used to "vacuum-pack" food in plastic—only this pump would be stronger, able to get more of the gas out). That would likely solve the condensation problem: your former ice cubes are now a gas. What is the shape of the water vapor? It takes the shape of the entire jar, not just the bottom of the jar the way the liquid did. What is the volume of the water vapor? The volume of the jar. The molecules of gas are barely held together at all.

WONDEROUS WATER

If you melt ice, the volume of liquid water is about 10 percent smaller than the volume of ice made from the same amount of H_2O molecules. We're used to that, but from a materials-science point of view, it is bizarre.

For just about all elements and molecules, the solid has less volume than the liquid state. (Oddly enough, silicon also expands when it solidifies, but not nearly as much as water.) The reason is that when water turns to ice, the molecules try to line up in the way that uses the lowest possible energy. The configuration they end up in—usually—is a kind of hexagonal structure. Unlike most solids, it is less space-efficient than the liquid form.

This causes a lot of weird behavior. One example: for most atoms and molecules, the solid form is denser than the liquid form, so the solids sink to the bottom in a mixture of the two; but for H_2O, the solid is less dense and floats on the liquid. Even really huge chunks of solid, like icebergs, float.

Another example: sometimes water soaks into cracks in rock. (That's not the weird part.) If the temperature dips far enough, that water solidifies. (That's not the weird part either.) The freezing water expands and pushes against the stone around it. And it pushes so hard that it can crack the rock. So in temperate areas, water becomes a major force for erosion even away from running streams and rivers.

These properties hold for all materials, not just water. Gases have the most kinetic energy, and the lowest density. Atoms or molecules in a gas whiz around, only occasionally bouncing off another atom. The molecules are not arranged in any particular order relative to one another. The air we breathe (which is mostly made of nitrogen) is a gas. Liquids (all of which have molecules with less kinetic energy than they would as a gas), stick to one another and conform to the shape of their containers, but can have free sides. (The sides that don't stick to a solid often hold themselves together with surface tension, which is another very cool effect.) Molecules and atoms in solid form have the least kinetic energy and the least freedom to move around: solids maintain their shape and are usually denser than the other forms.

The changes from one state of matter to another aren't always that simple. For one thing, the states mix. To use just one more kitchen example: there are solids suspended in the liquid of your hot cocoa, and gas inside solid bubbles in the whipped cream on top.

The discussion about ice described how temperature changes the state of matter, but it isn't the only thing that can cause a change of state. Changes in pressure can also change the state of matter, although this isn't likely to happen in your kitchen unless you use a pressure cooker. There is a certain combination of temperature and pressure at which a material can exist as a solid, or a liquid, or a gas. This funky combination of conditions is called the triple point, and every material has one. So if conditions are just right, gases can solidify without becoming liquids first, and vice versa.

A slight nudge, a tiny addition or subtraction of kinetic energy, or a minuscule change in pressure, and the material will change state. One can cool liquids well below their usual solidifying temperature by manipulating the pressure. (Although the liquid becomes very sensitive to the least

SEA LEVEL ON MARS

For pure water, the combination that defines the triple point is used as a definition. This turned out to be useful for exploration of the red planet. Like Earth, the surface of Mars is bumpy. How tall, and how low are its ridges? Where do you measure from?

We don't have that problem on Earth, because much of Earth is conveniently covered in water. This gives us a globally available low point from which to measure elevation. We can say that the top of Mount Everest is more than eight thousand feet above sea level. We can say that the Marianas Trench is even farther below sea level (nearly seven miles deep). Even when we talk about more moderate points like, say, the Ninth Ward of New Orleans, we can use sea level as a reference.

But Mars is awfully dry. If you don't have a sea, how can you have a sea level? Still, it would be convenient to have a reference for elevation on Mars. After some discussion, space scientists agreed to use the triple point of water to determine an artificial sea level on a planet that doesn't appear to have any surface water. If there were water on the surface, then the combination of temperature and pressure (from the planet's gravity) would be at the triple point at "sea level."

perturbation. Supercooled water, for example, will solidify if the surface of its container is rough, or if a pulse of sound goes through it.)

Chemists make state diagrams, with temperature along one axis and pressure along the other, and the area is divided into at least three regions, one for each state. Depending on where on the chart you cross from one state to another—especially for solids—you end up with different atomic arrangements.

Getting back to more earthly concerns: pressure can also change the state of matter. If you raise the pressure on a gas by squishing it into a small enough space, the gas will turn into a liquid. There's a lovely simple equation called the ideal gas law that describes how the temperature and pressure are related to the volume. Van der Waals equation is a little more accurate, since it also takes into consideration the size of the molecules and the forces between unbound molecules, sometimes called van der Waals force. (The Dutch physicist Johannes Diderik van der Waals really dug into the changes in gases and liquids when one changed their temperature and pressure. He won the Nobel Prize in physics for it, in 1910.) The van der Waals force is a force between molecules, which we usually don't notice because stronger forces overwhelm it. But it's cool because sometimes it pulls molecules together and sometimes it pushes them apart, depending on how far apart the molecules are. But explaining why that is, in detail, is a story for a chemistry or thermodynamics class.

GREAT GUMMY GECKOS STICK TO ANYTHING

Fantastic human rock climbers are sometimes called "human flies," but when it comes to sticking to surfaces, even flies are slouches compared to geckos. These neat little reptiles can climb walls and stick to ceilings no matter how slick—they can even climb polished glass.

They don't appear to use other common methods of climbing: human climbers depend on friction, but that wouldn't work for climbing on ceilings. Some other animals use glues of some sort, but the gecko doesn't produce adhesives or leave behind any residue, which you'd expect from a glue-using critter. They don't have suction cups on their feet and, in fact, can climb in vacuum. And although it seemed possible for a while that they used the surface tension of water to stick, in the way that frogs do, they don't appear to be using liquid at all.

So how the heck do they do it? They use van der Waals forces. Specifically, when two molecules are very close together, they are attracted to each other. Geckos' feet are covered with tiny hairs—and the tip of each hair is further subdivided into a sort of brush of even tinier hairs. (Split ends are a good look on geckos!) With such tiny hairs, and the way the gecko walks, the molecules in the hairs get so close to the molecules in the surface that they attract one another. (There is one surface that geckos can't stick to, though: Teflon.)

A whole host of researchers are looking for ways to copy gecko feet using nanostructured materials. And just in 2007, the usually staid *Journal of Physics: Condensed Matter* published a paper about how to make a Spider-Man suit. As we know from the comics and movies, Spider-Man does whatever a spider can do and a few things that a spider can't. He can spin webs and cling to surfaces. Italian engineer Nicola M. Pugnoc of the Politecnico di Torino analyzed new materials like the gecko-based tape and carbon-nanotube-based cables that are almost good enough to make a real Spider-Man suit.

Pugnoc suggests that window washers, among others, would benefit from wearing geckolike gloves that could stick to glass (or any other building surface). Personally, I think they'd make games of tag and hide-and-seek a whole lot more interesting.

FURTHER READING

Nicola M. Pugnoc, "Towards a Spider-Man suit: Large Invisible Cables and Self-Cleaning Releasable Superadhesive Materials," *Journal of Physics: Condensed Matter* 19 (August 2007).

Kellar Autumn, "How Gecko Toes Stick," *American Scientist* 94 (March–April 2006), 124.

Why do we bounce off walls?

So, given what we know about the states of matter, it's fairly clear why people (who have solid skins, even if we are made mostly of liquids) can walk through air: the molecules in air are sparse, and not arranged in a way that would provide much resistance. We can just push them out of the way.

We can even walk through water, although the resistance of a liquid is higher than that of a gas, and thus it's often easier to swim. (This is why some people with joint problems do water aerobics and water walking. They get extra resistance from the water, but because the water supports them more than air, walking in water puts less pressure on their joints.) In liquids, the molecules stick together, but they can slide past one another fairly easily.

Solids, though, are a different matter. Those atoms are stuck, joined at a molecular level. Going through something solid usually requires breaking the chemical bonds between molecules. Going though a solid wall requires either a tank that can overpower enough links to take a chunk out of the wall, or a very clever device that temporarily moves a chunk of solid out of the way: like a door on hinges. (Hinges really are one of the spiffiest simple devices around, ranking right up there with Velcro. Some of my other favorite inventions are slightly more complicated: woven cloth, modern plumbing, and fermentation are also marvelous.)

But what, on a molecular level, is D.L. doing when he walks through a wall? He is, somehow, getting the molecules in the solid to act like molecules in a gas while he pushes through and then resolidifying the wall behind him. (That last part is important: after all, superheroes like Superman and the Hulk can also go through walls. But they leave a superhero-shaped hole behind.)

Even if D.L. has a handy way of turning solids to gases, turning a gas back into a solid that resembles its previous state is quite a trick. Just consider D.L. going through a wall in the standard construction of his house. First, he goes through the paint, then through wallboard. Then parts of him go through the studs while the rest of him doesn't. Then he passes back out through another level of wallboard and paint. Just keeping track of where to reassemble layers without mixing them up is quite a trick!

That would, however, explain the occasional ripples we see when he pushes a hand through something that looks solid. What can change a state of matter? We've mentioned heat and pressure, but really, all it requires is getting the matter to absorb a lot of energy.

Liquefy your assets

Remember the short-lived Zane Taylor, who could turn solid objects into liquids (at least, until he met Sylar and lost his head)? That's really not such a bizarre trick for a superhero: Superman's heat vision, or maybe Cyclops' laser vision could achieve the same effect.

Zane was able somehow to convince the molecules of solids to move around more in relation to one another. If he deposited heat into the solids, it didn't seem to heat the surroundings. People who use lasers for machining and welding deal with this often, trying to minimize the size of something called the heat-affected zone. Surgeons using lasers face the same problem: sometimes they want to cut and cauterize tissue, but cooking the patient would defeat the purpose. If you deposit energy (by laser or any other method) into some atoms quickly—like, in thousandths or millionths of a second—then the area is heated before much heat can spread to the neighboring atoms. This doesn't keep the material from resolidifying afterward, but it's unclear whether Zane—and later Sylar—permanently liquefied objects.

One other concern is that depositing that much energy into something that fast can cause it to ablate. Ablation is a fancy

MICROSTRUCTURE MATTERS

In the movies, Superman has a great party trick: he grips a lump of coal in his fist and squeezes with his super strength until he ends up with a tiny sparkling diamond. Would that work? It depends both on the coal and on what you replace Superman's hand with.

Even if turning coal to diamond seems like turning lead to gold, it's not really. Lead and gold are different elements. But coal and diamond, in their purest forms, are both made of carbon. So how can coal be soft enough for artists to use when drawing on paper, while diamond can cut glass and steel? Also, how can things that are such different colors be made of the same atoms? Coal is, excuse the expression, as black as coal. But diamond is transparent.

Both are allotropes of carbon. Made of the same thing, the arrangement of their atoms determines many of their properties. Actually, coal tends to contain a lot of noncarbon atoms, even though they decrease as it goes through the (no kidding) "coalification process" from peat to anthracite. Graphite can be considered a superhigh grade of coal. The carbon atoms in graphite bond mostly in hexagonal arrangements in flat sheets, with relatively few, relatively weak bonds outside of the sheet. Therefore, the sheets can break apart and slip past one another easily. That's great for pencils, which use graphite as the "lead." It also makes graphite a great lubricant: it is slippery. Within the sheet, though, the bonds are very strong. In fact, sheets are rolled up to create very strong reinforcing fibers for composite materials like, for example, high-tech racing bicycles.

Diamond is one of the hardest substances known to man. In molecular terms, this means that it is difficult to break the bonds between the atoms. These atoms are not arranged in sheets. They are arranged in tetrahedrons (role-playing gamers will recognize this as the shape of four-sided dice), with each carbon atom connected to four others.

If you push the sheets of graphite together with enough pressure and heat, the bonds rearrange into tetrahedrons and hey, presto: diamond! Instead of squeezing with your fist, though, you'll need a high-pressure press: synthetic diamonds for industrial use are made like this.

FURTHER READING

Robert M. Hazen, *The Diamond Makers*, Cambridge University Press, 1999.
Amanda Yarnell, "The Many Facets of Man-Made Diamonds," *Chemical and Engineering News* 82 (February 2, 2004), 26–31.

way of saying "blowing it up." On the other hand, if you get laser eye surgery to improve your vision, ablation is the desired result. Little bits of your cornea are vaporized, without damaging the rest of your eye beyond its ability to heal.

Sylar at one point liquefied a wrench. Assuming that it was made of hardened steel, what temperature would it have to be heated to in order to turn liquid? Depending on what kind of steel it was, maybe around 2,500 degrees Fahrenheit. That's a lot of energy!

That number is variable, though: some types of steel are much more heat resistant than others.

So how could D.L. resolidify the gas behind him without noticeably messing with the arrangments of the atoms? It's a conundrum.

Actually, there is another alternative: instead of changing the solids he encounters, D.L. could be changing the state of matter of his own body. Some things do pass through solids without too much trouble: consider light through solid glass windows and the cell phone signal that penetrates through the walls of your home. But even if D.L. becomes gaseous, some of the things he goes through are not easy for gas to pass through. Even molecules as small as helium take a while to leak through the latex of a balloon. And if, instead of a gas, he changes into something other than matter, then his ability is really more like Hiro's ability to teleport.

Quantum tunneling!

That might be it, exactly. Or, rather, maybe D.L. only teleports short distances through solids. There is precedent for that sort of behavior. It's called quantum tunneling, and it is one of those phenomena that doesn't seem to make sense on a human scale, but nevertheless has been predicted and observed on the scale of atoms and electrons.

Everything that we are used to thinking of as a particle also has a wave nature. This is most obvious with very small, very low-mass things. For example, light. We can talk about a particle, a quantum, of light, which is called a photon. Photons—not too surprisingly—travel at the speed of light, by definition. We think about photons as particles in order to describe how they bounce off mirrors and travel in straight lines. But they can also be described as waves of electromagnetic energy.

Occasionally, one needs to keep both models in mind at once, which is confusing. For example, separate one laser beam into two beams and shine them on the same wall. They will interfere with each other: sort of like the ripples from two stones tossed into a still pond will overlap, creating higher ripples in some places but canceling out each other's troughs and peaks in other places. The light on the wall shows an interference pattern. When you have two beams from a laser, with lots of photons, it's easy to think of the light as a combination of waves. But what happens when you shoot one, single photon? You still get an interference pattern. Sometimes the particle description doesn't work.

Electrons also have a wave nature. People have used the wave nature of electrons to make electron holograms, for example. Electrons have a wavelength smaller than photon's— a few nanometers, a few millionths of a millimeter.

A funny thing happens when something with a wave nature encounters a barrier. Most of the time it bounces off just like you'd expect a particle to. But some of the energy in the wave goes beyond the surface. For light, this is called the evanescent field, and it can be used to transfer light from one glass rod to another that is held very close to it. Even though light doesn't leak out of the first rod, there is a probability that the energy from the light will end up in the second rod—that some of the light will cross over into the second rod.

We've found all sorts of ways of using this property in electrons. In electrons, it's called tunneling, and we have scanning tunneling microscopes that image individual atoms on the surface of a solid by measuring how many electrons travel from a conductive surface across a vacuum (where they shouldn't exist) and into the conductive probe that gets very close to the surface.

An electron can also pass from one conductor to another one, through a thin layer of insulator. This is the basis of transistors in modern electronics. The tricky part is that it isn't leaking through the insulator, you can't trace the path that it followed, and there isn't any electron-size hole that it could pass through. Some of the electron's wave extends through the insulator. Occasionally the electron will disappear from the near side of the insulator and reappear on the far side.

The tunneling phenomenon isn't limited to photons and electrons. Atoms also have wavelengths, but they are even shorter than the nanometer or two of an electron since they are a lot more massive and thus have larger momentum. A person's wavelength is more like a trillionth of a trillionth of the width of an atom. And yet...and yet quantum tunneling might be the best explanation for how D.L. passes through walls.

So, say, D.L. misplaces his house key and he's in a hurry. He wants to walk through his front door. First, he gets close to the door. Then the molecules on the front of his body, acting as waves, have a little bit of their energy extending past the door into the house. D.L. somehow coordinates them all to locate themselves on the far side rather than on the near side of the door. As he's doing that, the next slice of himself, which was farther from the door, is now much closer to the door. Its wave nature extends a little ways through the door—repeat many times, very quickly, and what you have is D.L. inside the house.

Maybe D.L. somehow manages to make a Bose-Einstein condensate (as mentioned in an earlier chapter), which could coordinate all his molecules, get them all into the same quantum mechanical state, so he'd only have to switch his location from outside the house to inside the house in one tunneling event.

I'm not going to speculate about how he manages to bring his clothing with him.

A solid family man

Photons and electrons are one thing, but for something as big as D.L. to quantum tunnel is quite a trick! We don't know how to do that. We'd need a lot of advances in our ability to control all the quantum states of the molecules in our bodies before we could try it. Still, it is only exceedingly unlikely, not impossible.

In the end, D.L.'s heroism isn't a result of his ability to "phase," though. No matter what his talent in passing through solids, he's a hero because, when Niki's life is on the line, he chooses not to let the bullet pass through him, thus saving her life when Linderman shot at her.

Further reading

More about intermolecular (aka van der Waals) forces is available in a set of slides from Dr. Arlene Courtney of Western Oregon University. Available on the Web at http://www.wou.edu/las/physci/ch334/lecture/intermol/.

Anthony Hey and Patrick Walters, *The New Quantum Universe*, Cambridge University Press, 2003.

Chapter 8

MICAH SANDERS AND HANA GITELMAN GO BIOHACKING; ELLE BISHOP THROWS LIGHTNING

Young Micah Sanders is just learning to use his powers, whereas Hana Gitelman's are well developed by the time we and Ted meet her in "Unexpected." Nevertheless, both Micah and Hana are biohackers: Micah can talk to electronic machines and Hana can talk through wireless electronic networks. They can communicate without hardware.

We see Micah control several machines: he somehow convinces an out-of-order telephone to connect him to his mom's phone in "Nothing to Hide." And it requires only some thought and a touch for him to get money from an ATM a few episodes later. Micah is, pretty clearly, manipulating electronics. Moreover, he's manipulating them at the level of the machines—not at an interface designed to be easy for humans to use. And he manages it by touching the casing—which is grounded, so there is no direct electrical connection. Unlike Micah, Hana doesn't have to touch anything.

Micah and Hana use networks without hardware, sure. The really interesting question, though, is, when will we be able to do the same thing?

What are the networks?

In a way, our electronic networks are much like our nervous systems: distributed electrical signals passing along conductors. (That is a very limited simile: in many many ways, they are nothing at all alike.) But our nerves and brains are also chemical, analog living systems while our phone and Internet networks aren't.

When Hana meets Ted Sprague, she says she can receive satellite signals, wireless signals, and radio waves. And then she says a very odd thing indeed: she says she can "just pull e-mail out of your head." I hope that Hana was showing off and being imprecise, because there's a huge difference between being sensitive to intentionally transmitted communications at wavelengths of millimeters or longer (which are the wavelengths that our wireless technology uses) and being able to read minds!

So what, exactly, is Hana doing when she, for example,

intercepts and responds to a text message on Noah Bennet's cell phone? Normally, a cell phone communicates only with the transmitter on the nearest local tower. If Hana is horning in, that means that (from the cell phone's perspective) she is impersonating the tower.

First, she's somehow detecting a low-power signal on one of at least dozens and possibly hundreds of microwave frequencies at which a cell phone can transmit. She's also distinguishing Mr. Bennet's phone from any other phones in the area. And she's decoding the signal from his phone into Hana-intelligible text. (She doesn't necessarily have to decode it into normal-human-readable text. She's not a normal human, and if she can get the gist of the message without having to turn the signal into words on a display, it saves her a few steps.) There is plenty of information in the signal that she can just ignore. For example, she doesn't care at all about the carrier and billing.

Then Hana is thinking up a reply, reencoding it, and somehow transmitting it on a different wavelength—the one the cell tower uses to communicate to Mr. Bennet's cell phone—with enough power to be received by his phone. (Meanwhile, the real cell tower is still receiving the signal from Mr. Bennet's phone and sending it on to the cell phone company's local office. Unless Hana is also jamming that signal—which could be an entirely different skill.)

Tapping into the phone signal may be a lot harder than what I've just outlined, depending on the type of cell phone

Mr. Bennet uses. If the phone uses a spread spectrum system, the signal isn't sent over a single wavelength—it is split up and pieces of it are transmitted over a variety of wavelengths. This technology, first used for military applications, found its way into satellite communication systems, and is now also used by cordless phones and Wi-Fi networks (both of which work at longer frequencies than cell phones). When Hana communicates with Ted's laptop, she's definitely using spread spectrum technology. She's pretty agile, in a frequency-hopping sense.

Throughout most of human history, the ability to manipulate radio waves would have been useless. It is just possible that she might have heard the interstellar noise that has always been around, but wasn't noticed until astronomers started building radio telescopes. If Hana had been born even a hundred years ago, her ability would have been unnoticeable. She is very much a woman of our time.

It's not entirely implausible that Hana could have some extra sense that allows her to detect microwaves. Microwaves are electromagnetic radiation, which you can think of as photons that vibrate as they travel.

Are regular people or animals sensitive to electromagnetic radiation? Within certain boundaries, yes. We not only detect the presence of electromagnetic radiation at certain wavelengths, but our eyes form images with it. We call that the visible part of the spectrum. Remember how Claude's invisibility trick doesn't work on the guys wearing the night-

MAGNETIC FIELDS, ELECTRIC FIELDS, AND ELECTROMAGNETIC RADIATION

Most of us are familiar with magnets, and have experimental evidence that they create a field around them that can attract paper clips and sometimes repel other magnets. My son inherited one of my old toys, the one with iron filings and a plastic cover over a picture of a man's face: by using a magnet-tipped wand, we could position the iron filings to give him a beard, glasses, pointy hat, an education in graffiti, and a visceral understanding of magnetic fields.

A changing magnetic field creates an electric field. And, conversely, a changing electric field creates a magnetic field. We use the latter phenomenon every time we turn on an electric motor: the current creates a magnetic field that pushes or pulls at small magnets embedded in the motor, converting the electrical energy into mechanical motion. (We also use this phenomenon to create all manner of electromagnetic devices, such as the very powerful electromagnets in MRI machines.)

But really, electric and magnetic fields are different manifestations of electromagnetism, which is one of the four basic forces. And, in addition to the magnets we put on our refrigerators and the electricity we get from power outlets in our house, the rules of electromagnetism also apply to light.

Light is a wiggling wave with both electric and magnetic components. We can describe light in a few ways: by wavelength (the distance the wave covers before its wiggling begins to repeat itself), or its frequency (how often the wave repeats itself in a given distance, usually expressed as Hertz or sometimes as cm^{-1}), or occasionally by the energy it carries (usually expressed as electron volts). Any of these measurements can be converted into the others because they are all different ways to express the same thing. Radio waves, for example, have wavelengths that are meters long—in other words, they have low frequency, and they have low energy. Near the other end of the electromagnetic spectrum, X-rays have short wavelengths about the diameter of an atom, high frequencies, and high energy.

vision goggles that are sensitive to the infrared? They could see Claude and Peter's body heat. The infrared is the part of the spectrum that has wavelengths shorter than microwaves and wavelengths just a little bit longer than the color red, not to mention the rest of the visible spectrum.

Some creatures can see wavelengths that we can't: some fish can detect infrared and bees can see into the ultraviolet. Others are sensitive to magnetic or electric fields.

To a certain extent, we indirectly detect wavelengths that we can't see. When infrared sunlight falls on our skin, it is absorbed and converted into heat. (The sunlamps that some people have installed in their bathrooms work the same way.) And even electromagnetic radiation that we can't sense can still affect us, depending on the wavelengths, the intensity, and how much exposure we get.

For example, although infrared radiation may make your skin feel warm, the much shorter wavelength and more energetic ultraviolet radiation can burn your skin. If other types of electromagnetic radiation are absorbed, and therefore heat parts of your body, they could cause damage. There has been a lot of concern over the past few decades about whether power lines, cell phones, and the increasingly ubiquitous "electromagnetic smog" may cause or increase cancer or have other biological effects. But like many other environmental factors, proving clear cause and effect is difficult, and thus far we don't have scientific evidence to do so. So far, experiments have shown reproducible effects that are too small to consider dangerous.

THE HONEYBEE AND THE PLATYPUS

Some creatures navigate and hunt using magnetic or electric fields. Birds seem to sense the magnetic fields of the earth, which aids their migrations and accounts for the remarkable accuracy of homing pigeons. Honeybees also appear to use magnetic fields to navigate, and in experiments were trained to discriminate between relatively subtle differences in man-made magnetic fields.

These creatures contain a magnetic mineral called magnetite. Oddly enough, people also have a magnetite deposit in a bone in our noses. Whether the magnetite is necessary to sensing magnetic fields, and exactly what mechanism accounts for this sense, is still a source of debate.

Other animals—mostly sea creatures—can sense electric fields. Sharks, for example, use their ability to sense electric fields to locate prey right before they attack. .

But maybe weirder than sharks is the bill sense of the platypus. These Australian creatures are pretty odd mammals in the first place: although they are warm-blooded, hairy, and produce milk for their young, they also lay eggs. They look like a mishmash of other animals more familiar to North American eyes: they have the flat tail and pelt of a beaver, the webbed feet of an otter, and a bill like a duck. The males also have poisonous spurs on their hind feet. When European naturalists described them in the early 1800s, they caused a lot of confusion—first to decide whether they were a real creature or a hoax, then deciding just how to classify them.

But one platypusian ability that confused researchers for over a century afterward was how platypuses found the small mud-dwelling invertebrates (worms, etc.) that they eat. These creatures could close their eyes, ears, and nostrils, in the dead of night, and still catch the critters living in the bottom of pools and streams. That they were sensing in some way beyond the normal human senses was obvious. Figuring out what they were sensing—and how—was another matter.

The mystery was finally solved. In the early 1980s, someone noticed that parts of the platypus's skin resembled the electric field receptors in

...*continued from pg 181*

fish. Then, in 1986, a German-Australian team of researchers reported a pretty clever experiment that shows that the creatures do, indeed, sense electric fields. They hid live batteries in the mud. The platypuses found, and attacked, the batteries, as though they were prey.

Apparently your average platypus can sense its prey wriggling from the electric field generated by the prey's muscles!

FURTHER READING
Sönke Johnsen and Kenneth J. Lohmann, "The Physics and Neurobiology of Magnetoreception," *Nature Reviews Neuroscience* 6 (2005), 703–712.
John D. Pettigrew, "Electroreception in Monotremes," *Journal of Experimental Biology* 202 (1999), 1447–1454.

Hana, eventually, found herself riding a satellite down into the atmosphere, and undoubtedly died. But, it seems, there's a ghost in the machine, an intelligence in the network: after her fiery fall, another message was posted to her Web site. In the summer of 2007, Hana (or someone calling themselves Hana) compared the brain to a computer, and posited that the mind is like software, which could be installed on a new machine. The message said that somehow, someway, she was still with us.

This is puzzling (to say the least) because while we may occasionally compare software to our minds, and computers to our brains—our brains and minds really don't work that way. Although "uploading" memories or even personalities is nearly a cliché in science fiction ranging from *Star Trek* to

Doctor Who, we don't have that sort of technology—we're not even close, right now.

Human/machine interfaces

The Internet and wireless networks are fantastic tools for communicating—will we someday be able to use these tools without all the hardware associated with them? Maybe.

The first step is developing ways to interact with computers and machines that is easier for people. In the past thirty years, interfaces (as well as computing power and networking) have advanced ridiculously quickly.

A little over thirty years ago, when my oldest brother programmed a computer, he did so with punch cards: a stack of roughly three-by-eight-inch paper cards with holes in specific locations, to indicate different bits. The computer in question was a ridiculously expensive and large machine, a mainframe, available to limited users. Soon after, the interface to mainframes became the keyboard-and-monochrome monitor. This was such an improvement that—with the addition of a mouse and spiffier graphics—most of us use basically the same interface for our PCs today.

Twenty years ago, in order for me to send e-mail to a cute boy over the computer network, I had to go to the Computer User Center—a room full of workstations hooked up to the mainframe—at my university in Virginia, get special permission to access an intercollege network (Fidonet), address my message to a gateway computer (in New Jersey) that connected to a different network (ARPAnet), and then tell that network to send the message to a specific user (the cute boy) in Boston, who then had to go into a room full of workstations to read it. It was far more cumbersome than instant messaging, and an awkward way to conduct a romance. (It was worthwhile, though. Reader, I married him.)

Today, he and I sit at our laptops in different parts of our house, connected to the household's wireless LAN, and from there to the Internet. Still, despite stylus-controlled PDAs, the keyboard and mouse continue to be...if not the *best* interaction devices, at least consistently the least annoying.

How to make computers less annoying

Humanity did not evolve to type at a keyboard. In *2001: A Space Odyssey* (made in 1968), the Hal 9000 computer not only had artificial intelligence, but communicated by voice—and could read lips. Verbal communication is somewhat possible today, with programs like Dragon Naturally Speaking allowing users to speak to their computers, and plenty of programs that make the computer talk to us.

Some large companies use voice recognition systems on their automated phone systems. The ability of the system to recognize words in a variety of accents, over connections of varying quality, is pretty remarkable.

But voice-only systems can be frustrating. Why can't I point and say, "That one over there," and have the computer understand what I mean? New computer programs are beginning to respond: games like Dance Dance Revolution require you to move your feet, and the remote for the Nintendo Wii game system is designed to detect movement.

There's a class of potential computer users who cannot use keyboards: people who through disease or injury cannot control the muscles in their fingers. Some people can only move lips, or eyes, and struggle to communicate.

A camera that follows eye movements can allow users to move a mouse around a monitor. But why can't we just "jack in" to a computer, like Case in William Gibson's novel *Neuromancer* or Harper on the TV show *Andromeda*? Maybe someday, if implants can be made sufficiently biocompatible, we will.

Right now, and for the past five years or so, a number of research groups around the world have been researching brain–computer interfaces. The most obvious advantage of having such interfaces would be the way quality of life would improve for paraplegic patients—it could allow them to operate computers, wheelchairs, prosthetic limbs, and other devices, using brain signals only. But it could also allow people to control avatars in virtual realities, à la *The Matrix,* or

VIDEO GAMES IN YOUR HEAD

In 2006, the Berlin Brain Interface group—a collaboration of several European research groups—demonstrated a system that allowed users to play a very simple video game, Pong, controlling the cursor using only their brain activity. Pong is much like table tennis—a moving spot travels from one end of the screen to the other. Each player attempts to block the ball from exiting off the edge of the screen, using the paddle—a narrow rectangle. In this version of the game, the paddle only moved up and down (not side to side).

Before a game could start, however, each player had to be fitted with an EEG cap and led through about twenty minutes of training. Both the computer and the user were being trained. Because every individual thinks somewhat differently, the computer needed to sort data until it could recognize specific signals. Users were asked to think about moving either their right or left arm.

The important point, for this experiment, was that EEG electrodes on the skin can detect "motor intentions"—the electrical activity that goes on in your brain when you intend to move one arm or the other. The electrodes pick up this activity, amplify it, and send it to the computer, which transforms the signals into device control commands.

The goal is to allow users to move the paddle by thinking of moving their left or right arm. And, because people learn whether you really want them to or not, and the program offers feedback, the user adjusts somewhat to give the most effective feedback.

The Pong game was just one demonstation of the system. Using the same left-arm or right-arm method, it was also possible—although cumbersome—to choose letters and type words in a word processor with a graphic interface.

A few groups have tried using EEGs and at least one American company keeps promising to release a gaming interface based on it.

FURTHER READING
A description of the project, along with photos of the funny EEG caps, is online at http://ida.first.fhg.de/projects/bci/bbci_official/index_en.html.

better control robots in hazardous environments, or shave a few fractions of a second from the response time of fighter pilots, or race-car drivers, or computer gamers.

There are a couple methods that can be used to detect activity from the brain. One that is both noninvasive and can be taken out of the lab is based on electroencephalographic (EEG) recordings. A number of electrodes are placed on the scalp and measure the electrical activity of nearby neurons. The signals are small, so signal processing is necessary to separate signal from noise.

EEGs are not the only method for gauging brain activity, but they are one of the most portable and least invasive. Other types of brain-activity sensing include positron emission tomography (PET) scans and functional magnetic resonance imaging (fMRI) scans. These require large expensive machines, however, so they have tied users to labs. At least thus far they have. A new advance in magnets, however, has led to the development of MRI scanning with ultrasensitive magnetic sensors, which has allowed much lower-power (and smaller) magnets to be used. We'll revisit that in the chapter about Matt Parkman.

Even with training methods, however, today's brain–computer interfaces have been unable to work with roughly 20 percent of people who take part in experiments. We have a lot to learn about how to adapt computers to work with humans.

Sing the body electric

Compared to Micah and Hana, who manipulate electronics with relatively tiny charges, Elle Bishop is a walking power plant. When we first meet her in the episode "Fight or Flight," she kicks up a few sparks running her hand along the wall of a shipping container. Later in the episode, she kills Ricky by, apparently, shooting a bolt of lighting at him.

Elle generating the electricity isn't that much of a puzzle. You and I can generate electricity on dry winter days by rubbing our feet on a rug. Or we can use a wall plug to charge a capacitor. (A capacitor is an electrical device that consists of two plates near, but not in contact with, each other. It can temporarily store electrical charge, so devices or electrical circuits that need to build up a high voltage may use them.) Those of us with computers use wall plugs to charge the capacitors in our computer power supplies. Elle's superpower appears to let her charge up at will, and discharge with a fair bit of precision in direction and timing. The way that she discharges the electricity is pretty interesting.

The thing about electricity is that it wants to reach equilibrium. Positive and negative charges "want" to find each other, and snuggle up together in a nice electrically neutral pool. And that's pretty much the function of the ground. The ground, in electrical terms, is literally the ground, the earth, under our feet. We live on a planet that is more or less insulated from the rest of the universe by vacuum. (Although some electrical charges come visiting, mostly from the sun.) My favorite electrical engineer and electrician, Regis Donovan, tells me, "In the world of frictionless surfaces and ideal gases, 'ground' is just zero voltage." Of course we don't live in a world of Platonic ideals. It's messy, and includes bizarre phenomena like ball lightning, which we still don't know how to explain. Except that it floats, and may or may not be entirely electrical, and seems to resist grounding for up to a few minutes.

If Elle can generate a large voltage, she must somehow keep it from draining out to the ground while she's storing it, and when she discharges it, the most obvious place for the arc to terminate would be either the ground, or something that is conductively attached to the ground. But that wouldn't be much fun as a superpower—she'd have to touch someone to discharge. Getting the arc to zap where she aims it is much more useful. My guess is that she's somehow ionizing the air along the path she wants the electricity to take. Here's why:

The difference between a bolt of lightning, the spark in a spark plug, and the shock you get in the winter from static electricity is mostly a matter of magnitude. In all three, a voltage difference between two locations is large enough to allow a spark to travel through the air. (Sparks can also occur in vacuum—old radios with vacuum tubes worked this way.) Dry air is not terribly conductive; humid air is only slightly more so. (The resistance of air to conducting electricity—that is, its electrical resistivity—is more than a million ohms per meter.) But if the voltage is high enough, I mean, really high, roughly thirty thousand volts per centimeter, then the air molecules break into ions, which are much more conductive.

So: creating a conductive path from her own personal capacitor to her victim is sort of like someone shooting a Taser at the person. A Taser shoots out two wires, with barbs on the end, and then applies a voltage difference across them. Elle creates one ionized path—like a single Taser wire—and then allows the electrical current to follow that path. Instead of a Taser wire, she lets the ground be the other conductor.

This is not unprecedented. Some early telegraph systems used a single wire and the earth instead of two wires to send signals. Both operators would have metal rods banged into the dirt to provide a ground connection. If the connection wasn't working well, they'd sometimes drench the ground rods with water.

Which leads to the question, what was really happening in the episode "Cautionary Tales" when Elle's power was disabled by getting her wet and putting her feet in a tub of water? She was grounded before, so the tub of water doesn't make a difference. Being wet didn't stop her from generating a charge, but it seemed to stop her from discharging it the way she intended to. Maybe her superpower allows her some way of controlling the resistivity of her skin, and that's spoiled by water on her? But then, if someone wanted to attack her, they'd just wait until she was bathing or at the pool. And if water on the skin causes her problems with discharging, does this mean that she can't use her power in the rain? (That might be a good enough reason for Mohinder to move to Seattle!)

The speed of the future

Thirty years ago, the transistor had already provided plenty of benefits, but most people couldn't imagine a world in which cell phones and the Internet allow us to communicate as easily with people on the other side of the globe as with our neighbors, a world where a person can record a video on a tiny cell phone and share it with hundreds of people.

Micah and Hana can talk to machines, and talk through machines, without needing any hardware at all. The rest of us aren't all that far away—the difference between us and them, however, is that the computers will work with *us,* rather than requiring a superpower for us to work with *them.* Today we talk to one another and browse the Web on increasingly tiny cell phones, we have Bluetooth ear sets that allow us to talk handsfree—at the speed that technology is moving, it may not be too long before we can implant the electronics under our skin, and easily control our computers with just our voices.

Chapter 9

RADIOACTIVE TED

Life is tough enough for Peter Parker, bitten by a radioactive spider, but what about poor Ted Sprague, who is, in and of himself, radioactive? Ted's unpracticed abilities accidentally caused his wife's incurable cancer, and his inability to control his talent later blasts, burns, and crisps the Bennet house and Claire. Yet, still later, he uses his power to jam electronics and thus escape from the shadowy organization that runs the paper company.

Ted Sprague is driven by grief, by the need for revenge, and by a desire to lash out at the people he believes forced his power on him. Unfortunately—both for him and the world— Ted possesses one of the most destructive powers we've seen yet: he can emit radioactivity.

What is radioactivity?

But what the heck is Ted actually doing when he's zapping radiation at someone or doing his own equivalent of twiddling his thumbs by creating tiny fireballs? What the heck, in fact, is radioactivity? To understand the answer, let's look at some chemical history.

In the 1860s, Dmitry Mendeleyev organized a periodic table of the elements. An element is a form of matter that cannot be subdivided into other elements. You could think of elements as the physical equivalent of prime numbers. Just as a prime is a whole number that cannot be evenly divided by another whole number other than one and itself, an element cannot be divided into any other elements. So, for example, bronze is not an element because it can be divided into copper and tin. But copper can't be broken down into another element.

The real world is a big messy stew of solid, liquid, and gaseous elements, and each element has a distinctive personality: some, like the noble gases, are self-sufficient and stalwart, resisting attempts by other elements to join with them. Other elements are discontent to wander singly and when they get a chance they take the opportunity to cuddle up with one of their favorite other elements and form a happy partnership in a compound. Still others, like oxygen, are so sociable that they aren't too picky about whose company they're in, as long as they aren't alone. (That's a lot of anthropomorphizing. Sorry.)

A lot of chemical experimenting went into separating one

element from anything else, and figuring out what was an element as opposed to an alloy, or a compound, or a mixture, and so on. A lot of theorizing was also involved as people tried to impose some sort of organizing scheme on the fragmentary and bewildering information available about different elements.

This is where Mendeleyev comes in. An international gathering of chemists had agreed that the atomic weight of an element was important. But he went one step further: he organized the sixty-four elements that he knew about by their atomic weight, and saw a pattern. He suggested that the atomic weight of an element could be used to predict some of the chemical and physical properties of elements.

Although he didn't know it at the time, the atomic weight is made up of the mass of protons and neutrons in the nucleus of the atom (and also electrons, but in comparison with protons and neutrons, they are so light as to be negligible.) That use of atomic weight was later abandoned when it became apparent that the atomic number (in other words, the number of protons—which are usually balanced by an equal number of electrons) in an atom gave a more accurate method for organizing the elements into groups.

This research was going along very well, and a number of new elements were found based on the predictions of the periodic table. And then radioactive elements came to the attention of researchers. The early researchers, in the late nineteenth century, included Henri Becquerel, Marie Curie and her husband, Pierre, as well as, later, Ernest Rutherford.

Radioactive elements were weird, even compared to the wide range of quirks of other elements. Radioactive atoms emit three different kinds of radiation. All radioactive elements emitted energy in the form of alpha particles, beta particles, and gamma rays. (If this sounds arcane and mysterious, it might help to remember that alpha, beta, and gamma are the first three letters of the Greek alphabet. There's no special significance to the names, except that the early researchers had no idea what to call the emanations when they first started research.)

The alpha particle has a positive electric charge and consists of two protons and two neutrons. In other words, an alpha particle is a helium atom that has been stripped of its two electrons.

Radioactive elements also spit out electrons, sometimes called beta particles, which are (by comparison to alpha particles) very light and have a negative charge.

Gamma waves, the third type of radiation, are electromagnetic waves—just like radio waves and light and X-rays, only with much much shorter wavelengths and (thus) higher energy. Cosmic rays are mostly gamma waves.

It is also possible for the nucleus of an atom to split entirely when bombarded with other particles—fission. This is the process used in nuclear power plants. This is also a good reason why Ted shouldn't be around materials that can split relatively easily.

In the late nineteenth century, research into radioactive

materials was really, really cool—it had the equivalent of the glamour surrounding today's nanotechnology and proteomics. Here's why: even as the periodic table was providing insight into the families of elements, the research on radioactive elements was breaking down the basic idea that elements were indivisible units of matter. Until then, scientists assumed that the element was the smallest possible unit, that it couldn't be broken down further, and that once a particular element, always that element from the creation of the universe until the end of time. But the process of radioactive decay is the process of transmutation, the conversion of one element into another, the long-standing dream of medieval alchemists who tried to turn lead into gold.

BOB THE ALCHEMIST

According to the medieval alchemists, a mysterious substance called the philosopher's stone would act as a catalyst, allowing lead to be turned into gold. (Yes, this is the same philosopher's stone that we read about in the first *Harry Potter* book. When the American publisher changed the title to *Harry Potter and the Sorcerer's Stone*, the term was stripped of hundreds of years of cultural history.)

Alchemists never succeeded for a simple reason. No catalyst would be sufficient for the transformation they desired. They were using chemical reactions and this conversion requires a nuclear reaction.

Lead, by the way, has an atomic number of 82. Gold has an atomic number of 79. The odd number of protons that is the difference

. . . *continued from pg 199*

between the two elements means that you can't turn one into the other using only alpha particle decay. If, instead of lead, you start with bismuth, which has an atomic number of 83, the conversion is slightly more likely.

According to an anecdote, Nobel Prize winner Glenn T. Seaborg managed the bismuth-to-gold trick in 1980 while at Lawrence Berkeley Laboratory. It was a ridiculously expensive method of creating gold, but it seemed like a worthwhile enterprise to the man who discovered or codiscovered ten elements.

On *Heroes,* Bob funds the Company with his talent for turning anything into gold. This also requires nuclear reactions—either adding or subtracting protons from the nuclei of atoms. Had Ted survived long enough to learn it, he might have been able to create gold, too. Let's hope that Sylar doesn't figure it out!

The Rutherford model of the atom is not quite correct but is sufficient for our purposes at the moment. This model suggests that every atom contains a nucleus composed of neutrons, which have no electrical charge, and positively charged protons. A swarm of negatively charged electrons outside the nucleus offset the charges of the protons. The electrons (which Rutherford thought were organized into shells, and later models determined were organized into orbitals) mostly determine how the atom interacts with other matter—or, more concisely: electrons give the element its chemical properties.

Elements are determined by the number of protons in the atom, but sometimes the number of neutrons and electrons vary. If the electrons vary, we may call the atom an ion. If the number of neutrons varies, then we have an isotope. Because the neutrons weigh a fair bit, this means that different isotopes of the same element have different atomic weights.

Some isotopes are quite stable. Not all isotopes are radioactive; that is, not all emit a radiation. Those that do are sometimes called radioisotopes.

For example, the lightest element, hydrogen, has the atomic number 1. Hydrogen is usually composed of one proton, no neutrons, and one electron, with an atomic mass of 1. But hydrogen has isotopes with different atomic masses. Deuterium is one isotope: it still has one proton and one electron, but also has a neutron, and an atomic mass of 2. Tritium has one proton and one electron, but two neutrons, and thus its atomic mass is 3. Because all three forms of hydrogen have the same number of electrons, they share the same chemical properties. Tritium, by the way, is radioactive.

Transmutation happens when an isotope emits alpha, beta, or gamma radiation. But for simplicity's sake, let's just consider what happens when alpha particles are emitted: suppose you were to get ahold of some enriched uranium. (*Don't try this at home!* You might find that you have many very concerned and very serious people knocking at your door if you do.) Enriched uranium is also known as uranium-238.

It is an isotope of uranium that contains 92 protons and 146 neutrons. And suppose an atom of this uranium emits an alpha particle. What's left is a nucleus with 90 protons and 144 neutrons. If it has 90 protons, it is no longer uranium (atomic number 92), but thorium (atomic number 90). It is, however, not a stable form of that element—it is an isotope of thorium that is also radioactive and likely to emit another alpha particle, thus reducing itself to yet another element. This chain of emissions and transmutations continues, drifting down the periodic table, until a stable atomic nucleus forms. In this example, that occurs when the atom has lost all but 82 protons—in other words, when the atom has become lead.

So that is, roughly, *how* the radioactive decay process happens. But it doesn't tell us anything about *when* an alpha particle will be emitted. We don't know, for sure, when that'll happen. The best we can do is figure out the half-life. Half-life sounds nonsensical (after all, people and things tend to be alive or dead, and any lifetime tends to have a specific beginning and a specific end), but it is a useful concept.

Suppose we have a hundred atoms of the same radioactive element. At some point, fifty of them will have decayed, and fifty will not. That point in time is a good first approximation of the half-life. If you do this experiment lots of times, enough to weed out the statistical improbabilities, you'll have a better estimate of the radioactive isotope's half-life.

Transmutation can be a pain in the rear if you're a chemist, as was Madame Marie Curie. She wanted to experiment with polonium, but had a heck of a time trying to find it and separate it from other elements. The longest-lived isotope of polonium had a half-life of only 138 days, which meant that even as she meticulously attempted to isolate it, the isotopes were emitting alpha particles and decaying into other elements. Curie is more famous for isolating and identifying radium. Its half-life of sixteen hundred years was long enough to avoid the frustrating decay problem of polonium but short enough to still show a lot of detectible radioactivity.

Uses of radioactivity

For scientific purposes, a great way of deciphering the atomic structure of unknown materials involves bombarding it with known particles, and looking at where those particles end up. Radioisotopes offered a source of high-energy particles, which were used until particle accelerators were invented in the 1930s.

And maybe you've heard of dinosaur bones or other artifacts from the far past being subjected to carbon-14 dating? Carbon 14 is an isotope of carbon, and since the 1950s radioactive carbon has been used to pin down the age of plant and animal remains.

What does radioactivity do to us?

But what does radiation do to us? It depends on what type you mean. Alpha particles exit radioactive atoms with high energies, but they lose this energy as they move through matter. An alpha particle can pass through a thin sheet of aluminum foil, but it is stopped by anything thicker. In contrast, beta particles travel at very nearly the speed of light and can make their way through half a centimeter of aluminum. Gamma waves, also called gamma rays, emitted by radioactive atoms can penetrate deeper into matter than alpha or beta particles. A small fraction of gamma rays can pass through even a meter of concrete.

All radioactive emissions are dangerous to living things because they all have the capacity to ionize—that is, to knock an electron out of—atoms, which plays merry hell with the delicate electrical balance of atoms organized into living tissue. Alpha particles, beta particles, neutrons, gamma rays, and cosmic rays are all ionizing radiation, meaning that when these rays interact with an atom, they can knock off an orbital electron. The loss of an electron in an atom in your DNA can cause genetic mutations that (if not repaired and paired with accumulated damage) can lead to cancer. Our cells contain a number of repair mechanisms for DNA, but they also are programmed to self-destruct if the damage is bad enough.

Because alpha particles are large, they cannot penetrate

very far into matter. They cannot penetrate a sheet of paper, for example, so when they are outside the body, they have no effect on people. If you eat or inhale atoms that emit alpha particles, however, the alpha particles can cause quite a bit of damage inside your body.

Beta particles penetrate a bit more deeply, but again are only dangerous if eaten or inhaled. Gamma rays, like X-rays, are stopped by lead.

Medicine

On the other hand, sometimes radioactive materials can be used on people for good, humane causes. Radioisotopes that decay in a few minutes or hours are used for a variety of imaging tests because they give off radiation that can pass through body tissues. Their use can result in lower radiation levels than using an X-ray would do. They are used for PET (positron emission tomography) scanning, imaging blood traveling through the heart, and imaging metabolic hotspots in bones.

When I say short-lived, I mean it: PET scanning typically uses isotopes like carbon 11 (which has only about a 20-minute half-life), nitrogen 13 (about 10 minutes), oxygen 15 (a mere 2 minutes), and fluorine 18 (about 110 minutes). Because they decay so rapidly, the hospital needs to generate these radioisotopes nearby. They are typically created with a cyclotron, a type of particle accelerator. Cyclotrons can also be used to irradiate tumors.

Sometimes exposing someone with cancer to radiation can be more effective and safer than surgery or chemotherapy for attacking the diseased tissue. Radiation damages DNA. Cells have mechanisms for repairing damaged DNA, but for cells that are dividing quickly, as cancer cells tend to do, there just isn't time for repair: it is a case of killing the tumor before the tumor kills the patient. By focusing intense X-rays on the tumor, that area is exposed to a lot of radiation while the rest of the body is not. In some cases, a piece of radioactive material is placed near the tumor; for example, men with prostate cancer may be treated by putting a small "seed" of a radioactive material near the prostate.

Unfortunately, radiation also is more selectively deadly to other cells that grow quickly, like hair, skin, and blood cells, as well as the cells that line our guts. This helps explain why people undergoing treatment for cancer may have their hair fall out and feel sick to their stomachs.

How Ted manages to generate radioactivity is only one question. Another is how he survives it. Maybe he has a talent a little like Claire's healing ability, only specifically focused on repairing the damage to his DNA. Maybe he has supershielded DNA. All of us keep most of our DNA wound up into tight little bundles, like spools of thread, with only a little bit showing at any one time. To transcribe a section of DNA onto messenger RNA, one section gets pulled loose from the bundle, copied, then tidied away again. Maybe something in Ted's DNA, or in the nucleus around it, super-absorbs or deflects ionizing radiation.

Superabsorbent Ted?

Speaking of superabsorbency...When Ted was in the custody of the FBI in "Seven Minutes to Midnight," he was interviewed by Matt Parkman and his FBI partner. Both of them wore film dosimeters, which are simple devices that indicate how much radiation a person has been exposed to. The dosimeter that Matt wore changed color—from green to red, indicating dangerous levels of radioactivity. But then the dosimeter turned green again.

Dosimeters aren't like your car's tachyometer or gas gauge, which go up or down depending on the current status of the thing they measure. Instead, a dosimeter is more like your car's odometer: one measures the accumulated sum of radioactivity while the other measures the accumulated sum of miles. They are designed to stay the same or go up. Neither of them is designed to decrease.

Film dosimeters are—no surprise!—made of film that reacts when exposed to radiation. Just as you can't "unexpose" thirty-five-millimeter film in your camera, you can't "unexpose" film on a film dosimeter. So I'm going to take the color change as a narrative device, showing that Ted is not only capable of emitting radiation, but that he could also somehow absorb radiation from his surroundings. (I have no way of explaining this—the particles and rays generated by radioactive decay go in all directions.)

Ted died not long afterward, another victim of Sylar's murderous rampage. But if Ted did have the ability to absorb and destroy radioactivity, he could have been a boon to the nuclear industry. If he could absorb (and perhaps hasten) the radioactive emissions from nuclear waste, he could single-handedly overcome the great objection to nuclear power. He would have had a job for life.

Even better, Ted—unlike us—could walk into Chernobyl or into other areas too radioactive for normal humans to survive, and stop the damage caused by the nuclear disaster. Which reminds me: back in the late 1970s and early 1980s, a television show called *Mork & Mindy* starred Robin Williams as an alien called Mork. He was complacent about humans using fission to create energy until he learned that we didn't have "Nuke Away" to deal with nuclear waste. (Then he was horrified.)

Had Ted been given the chance, and had he agreed, he could have been our Nuke Away. Of course, using Ted as a human industrial waste disposer is no more ethical than using Claire as a universal organ donor. But maybe by studying Ted as he used his ability, we could come up with a more ethical way of dealing with nuclear waste.

There is never a free lunch—energy never comes from

nowhere. So where does the energy come from to produce emissions from radioactive atoms? It lies within them. There is vast energy stored within the nuclei of atoms—namely, the nuclear energy—much more than can be liberated by mere chemical reactions between elements. This is why nuclear power (both fission, as is practiced now, and fusion, which is in development) is so promising: it's not a free lunch, but it provides a way of exploiting the power within each atom for human good.

This is also why atomic bombs are so horrific. And why Ted (or Peter or Sylar), out of control, is a terrible accident waiting to happen.

Further reading

"Marie Curie and the Science of Radioactivity," AIP website: http://www.aip.org/history/curie/.

Craig C. Freudenrich, "How Nuclear Medicine Works," at http://science.howstuffworks.com/nuclear-medicine.htm.

UMich pages, including http://www.umich.edu/~radinfo/introduction/risk.htm.

Chapter 10

MATT, THE HAITIAN, MONICA: MINDS AND MEMORIES

Often, we wish our families and friends would hear what we mean, rather than what we say. Matt Parkman, on the other hand, doesn't only hear what his wife says, he hears what she doesn't say. This causes some problems, to say the least, with their marriage.

In a world of increasingly intrusive monitoring technology, of closed-circuit TV cameras on city streets, of paparazzi armed with cell phones, and of spy satellites, our privacy seems to be shrinking almost daily. We have not quite reached the Orwellian future of *1984,* with its constant surveillance, constant propaganda, and its thought police...but can we still depend on the privacy of our own minds?

Is mind reading possible? Well, what does mind reading really mean? Can we hear unspoken words that someone thinks, the way Matt hears thoughts? Not today, and it doesn't look terribly likely for tomorrow either. On the other hand, people do often think in words—or at least, language shapes the way people think.

Here's a different question—one more suited to the still-nameless *Heroes* character called the Haitian—can we somehow picture a memory that another person is recalling? Plenty of stories depend on memory or the lack of it, from the amnesia plot device used regularly on soap operas to *Total Recall*. Why can't we back up our memories to a separate device, like we do our hard drives? To a certain extent, we can do this by using the marvelous technology developed about five thousand years ago: written language. This is exactly the purpose of keeping a diary. Sharing one's memories via a memoir or journal or autobiography is the closest way we have to commune with the future, or share the thoughts of the dead, including Samuel Pepys, Matsuo Bashō, Henry David Thoreau, Anne Frank, Winston Churchill, Anaïs Nin, Helen Keller...

Before we can learn how to read minds, we would first need to learn how to read brains. And that is harder than you might expect.

Brain basics

When we talk about a book, there are (at least) two different discussions we can have. We can talk about who wrote it, is it a cookbook or a novel or a thesaurus? How well does the author's style fit her subject? All the things you learned in high school English class are about the content of the book. But books also have a physical existence: they are made of paper, cardboard, binding, ink. A thick leather-bound family Bible and a shiny paperback romance novel are different objects, but both are books. Occasionally the content and the physical form get separated: the contents of a book can be read aloud from a CD or podcast, or on an electronic reader, or (courtesy of Project Gutenberg and others) downloaded from the Internet. But usually, when we talk about books, we talk about the physical things made mostly of paper that carry ideas in their ink.

Our minds and brains are a little like that. Mind is what the brain "does." Our minds contain memories and thoughts, philosophies, desires, calculations. Our minds hold content, the precious cargo of much of our identity. Our brains, on the other hand, are the physical structure that makes it possible: our brains are the scaffolding that supports consciousness, thought, memory, and emotion. Our brains also control our breathing, receive and interpret sensory information, and transmit information—including commands—to muscles and other organs. Parts of our brains regulate the endocrine system and our hormones.

How a brain manages all this is still pretty mysterious. We know some things about how the brain is organized. We know that a human brain has a cerebrum, a cerebellum, and a brain stem. We know that a cerebral cortex is divided into right and left hemispheres, and those are divided into lobes. We know that our brains are highly individualized: a lot of the structure of our brains develops after birth. We know that the cerebral cortex's surface grows faster than the volume, creating wrinkles that vary from person to person.

We know that certain parts of our brains are associated with certain functions, including the different senses, use of language, libido, making choices, etc.

We also know some things about the components of our brains: they incorporate specialized cells called neurons. A *lot* of neurons: in her three-pound brain, the typical adult has something like a hundred billion neurons. (This is the same order of magnitude as the number of stars in our Milky Way galaxy. And it's hundreds of times larger than the number of transistors on even the best computer chips.)

There are something like fifty different types of neurons, each with multiple tendrils. But they all receive and send electrochemical messages with other neurons. An individual neuron sends out electrical impulses that travel fast (as fast as thought, or to be more specific, about 220 miles per hour) out to the tips of its tendrils. A material called myelin acts as an electrical insulator across parts of the neuron. But at the tips of the tendrils, the impulse sends chemical neurotransmitters to cross fluid-filled gaps, called synapses, to another neuron. (By the way, there's an old urban legend according to which we use only a fraction of our brains—that's not true. We don't use all of the neurons all the time, but we use a lot of them.)

Neurons aren't the only thing in our brains. About 90 percent of the brain is made of glial cells that support neurons—by mopping up leftover neurotransmitters, by forming myelin, by guiding new neurons to the appropriate places in developing brains. (Recent research suggests that glial cells do some communicating on their own. This could be another layer of communication and complication to the brain. We know very little about this—most brain biology research has focused on neurons.)

That sounds fairly straightforward, until you realize that each of the hundred billion neurons may have a hundred thousand connections to other neurons, that messages can arrive a thousand times a second, and that the neurons are constantly forming new connections and changing the strength of existing ones. Now add in the variability from person to person and the still-confusing actions of glial cells. The difficulty of decoding the mechanisms of how we think begins to make sense. No wonder we don't know what's going on in there!

That doesn't stop scientists from trying to figure it out, though.

Dead and the quick

Our first clues about how the brain works came from studying brains of dead people and from studying how damage to part of the brain alters the way people act. Animal studies, particularly with mice, have also helped us tease out specifics of brains. Some diseases, like Alzheimer's, still cannot be confirmed until after death, with a physical examination of the person's brain. But although these methods are still used, we now have some less invasive and noninvasive methods of examining what's going on in the brain. The great benefit of these methods is that they allow us to consider what goes on in a normally functioning human brain.

We can look at the electrical activity in neurons, using electroencephalography (EEG). Electrodes placed on the scalp measure the electrical activity of many neurons "firing" at once, which can be used to create maps showing which parts of the brain are most electrically active. Because EEGs can show changes as fast as a few milliseconds, they can also be used to show when the activity occurs with good precision (synchronization between different parts of the brain appears to be important) when a person is set to a mental task.

The brain–computer interface I mentioned in the chapter about Micah and Hana uses electrical activity to allow a person to play a simple computer game using only brain activity (plus training on their part and adaptation on the computer's part).

Could Matt be reading minds by sensing the electrical activity in them? That would be a bizarre ability for a human, but it isn't entirely ridiculous. A swimming platypus can sense even as little electrical activity as the wriggling muscles of a worm. Brain activity might not be that much harder to detect.

Distance could be a problem. When a platypus employs its "bill sense," it is in water—a much more conductive medium than air. On *Heroes,* though, in the episode "Run," Matt hears Niki's internal argument as she attempts to talk Jessica out of trying to kill Matt's client—at that point, he is on the far side of a wall and a shut door and at least a few yards away from her. His electrosense must be remarkably sensitive.

To be a human EEG detector, Matt would also have to be insensitive to or be able to ignore the electrical noise around him that is generated by computers, by the electrical circuits powering the building, by the security systems in the building, and so on. Matt hears only people's thoughts. If he were sensitive to it, even the low-power circuitry in a cell phone might be unbearably noisy to him.

I think therefore I—what?

But Matt hears only the words of people's thoughts. Do people think in words? In fact, we do. Not exclusively, but language makes a difference in the way we think, and the lack of a name for a concept makes thinking about it more difficult. Turning an original thought into spoken or written words can often be a struggle, forcing us to work at paring down the potential meanings to fit specific words—which is perhaps why so much classwork in school focuses on developing this skill—and language provides a tool for sharpening meanings. Language can also be a scaffold for extending thoughts with analogies and rhythm, alliteration and imagery, that can turn amorphous thoughts into poetry.

Language is also a skill unique to the human species. Other animals communicate—some even have fairly large vocabularies, like crows—but calling their communications a language is stretching the definition of a language.

New ways of seeing

Our science isn't as sophisticated as Matt is—we're still learning how to read brains. Another method of imaging also uses the electric currents in your brain. Magnetoencephalography (MEG) looks at the magnetic fields produced by the moving electrical charges in the neurons—it is similar to an EEG, in that sense. Actually measuring the very small fields requires remarkably sensitive detectors. These devices usually use superconducting detectors, which must be cooled to cryogenic temperatures.

SQUIDs ON THE BRAIN

Superconducting quantum interference devices (SQUIDs), such as those used for MEG, were also recently used to provide a brand new type of magnetic resonance imaging (MRI). Most MRI scanners use a huge magnet to align the hydrogen atoms (mostly in water molecules) in your body. Then the scanner sends out radio signals to jostle them out of alignment. A detector records the strength and location of the signal (the resonance) that the atoms emit when they realign themselves to the magnetic field—as well as how long the signal lasts.

MRI scanners typically require magnetic fields about ten thousand to a hundred thousand times stronger than Earth's magnetic field. Running the electromagnets is expensive, cumbersome (nearby electrical equipment has to be shielded from the magnet), and cannot be used on people with metal implants, because the magnetic field can cause the implant to move, or to heat up, and damage the tissue around it.

But Vadim Zotev and his coworkers at Los Alamos National Laboratory in New Mexico demonstrated that they could scan a brain using a much lower magnetic field that is about a hundred times weaker than standard machines, and then use seven of the exceedingly sensitive SQUIDs to pick up the return signals. The smaller magnets reduce the cost of running the machine. Also, the low-field MRI scanners can be more open than the claustrophobic tube required today. The first brain scan produced by these means was blurry compared to standard MRIs, but such machines could provide cheaper, safer, and more patient-friendly scans in the future.

While writing this book, I underwent two MRI scans, which involved entering a room with a huge magnet and lying very still for half an hour in a noisy environment. I made the mistake of bringing my wallet into the room instead of stowing it in a locker. Afterward, I had to replace all my credit cards, because the data strips were wiped by the magnetic field. I'd also had a flash drive in my pocket. To my great relief, the notes for this book that were stored on a flash drive survived intact!

Electrical activity isn't our only way of seeing brain activity, although it is the most direct. Thinking is hard, energy-intensive work. Although your brain only weighs about 2 percent of your total body weight, it uses up 20 percent of your body's fuel.

When part of your brain is active, it uses up energy. To replenish the supply, the blood flow to that area increases over the next second or two. We can infer brain activity by watching where blood flow increases. There is a short time lapse between activity and blood flow, but for some studies that doesn't matter (or it can be coupled with a faster-responding EEG). Two methods that have been in use for the last two decades are functional magnetic resonance imaging (fMRI) and positron emission tomography (PET).

Functional MRI (fMRI) is a variant of MRI that looks at changes on a short time scale. It can be used to look at oxygen levels in the blood. Red blood cells have different magnetic properties depending on whether or not they are carrying oxygen; these properties can be used to detect where they drop off their oxygen.

Diffusion tensor imaging (DTI) is also based on MRI. Because DTI shows the direction of water diffusion, it also shows some of the structure of long skinny tentacles (the axons) of neurons, and thus can be used to image the connections between neurons.

PET is an entirely different sort of imaging method. It involves injecting a person with small amounts of a radioactive isotope—a short-lived isotope with a half-life ranging from two minutes to two hours. The isotope is integrated into a molecule that the body uses—oxygen is one possibility, glucose is another. After injecting the person, the molecules are carried to the area of interest (such as the brain), and concentrate in the most active areas. When the isotope decays, it spits out positrons, which can be picked up by the detector. Lots of positrons from one area indicate a lot of activity in that area. The radiation dose is minimal (at least, compared to CT scans), but the technique is limited because the isotopes have to be made nearby. This technique can look at blood flow, oxygen or glucose metabolism, or the concentration of dopamine transporters—one of the chemicals that transports signals across synapses.

All these techniques are being used to reveal what parts of the brain do what, as well as to locate damage and, in some cases, to help doctors plan neurosurgery. Brain atlases are literally maps, charting out areas of the brain and what electrical activity is connected to what mind activity. So although we can't read specific thoughts, we do know which parts of the brain are associated with which functions. Examples include activity related to sensations, movement, libido, choices, regrets, intentions, and memory.

Brain scans can detect the intention to do something, like move a limb, before the action occurs. Some researchers, having found the parts of the brain that control movement in arms or legs, have begun using activity in those areas in order to control prosthetics for people who have lost limbs. The

fNIRS MEASURES FRUSTRATION

One problem with computerized teaching tools is that sometimes information is presented too slowly and bores the learner, while at other times the information can come too fast and overwhelm her. Wouldn't it be great if a computer program could alter its pace to fit the learner, like a good tutor?

A good tutor watches facial expressions and body language to gauge frustration, but how can a computer tell whether you are bored or overwhelmed? It can tell by looking at how much activity is occurring in your brain. Tufts University researchers, including professors Sergio Fantini and Robert Jacob, recently used an innovative brain imaging technique to measure how hard people were thinking.

Functional near-infrared spectroscopy (fNIRS), like an MRI, detects how much oxygen is being carried by blood in different parts of the brain, but it uses infrared light. Unlike visible light, which is mostly blocked by skin (not to mention bone), infrared light can penetrate skin, skull, and brain tissue. (Infrared light is also safer than X-rays: it carries less energy and thus poses less risk of damaging tissue.) By measuring how much light, and which wavelengths, reemerge from the person's head, doctors can tell whether blood in a particular spot is full of oxygen or not.

Unlike in MRI scans, the subject of a fNIRS scan can move around. Because the light used for the scan is delivered to a helmet using fiber optics rather than a magnetic field, the user can move while being scanned and doesn't have to lie in a tube surrounded by a huge machine. Because the equipment for fNIRS is portable, it could be used when the subject is in a more ordinary situation than she is during an MRI scan.

detectors aren't reading the user's mind, but they are reading a small piece of the user's brain.

Brain scans can also detect patterns of activity in parts of the brain used for adding numbers (which, it turns out, is different from the pattern of activity for subtracting numbers). And, most of the time, when subjects were asked to think of either a place or a person, the pattern of activity was different enough to be distinguished.

But thus far, no brain scan can tell that you are, for example, remembering your grandmother. And the parts of your brain that are active when you think of your grandmother may be very different from the way that someone else thinks about her grandmother, so no single "memories of grandmother" pattern of connections probably exists from person to person. Thus far, at least, our thoughts are really private.

Memories make the man

Another character on *Heroes* doesn't read minds—he erases parts of them. The Haitian can block memories. Without our memories, without remembering our histories and the things we have learned, our very identities are different as well as the decisions we make. We don't have freedom of thought when our memories are tampered with. Without freedom of thought, how can there be freedom to act? And without freedom to choose one's actions, one cannot choose to act heroically—or villainously.

But what are memories? We visited the concept in the chapter about Claire, but let's revisit it. A memory isn't a physical thing. It might be a pattern of excitation across the brain, or it might be a pattern of connections between neurons in the brain. Or it might be both.

At least we know that the Haitian isn't destroying memories. After the Haitian erased Peter's memories, young Mr. Petrelli recovered them. The Haitian's superpower is more like hiding the memories than destroying them. It may be a slight comfort to know that the person rifling one's mental filing cabinets and plucking out memories is merely misfiling them rather than irrevocably shredding them. How he manages this is another question. Maybe he's interfering with the neurotransmitter chemicals that travel across our synapses? Maybe he can overwrite certain memories? We don't know.

We store two different types of memories, apparently in two different systems. The first, declarative memories, includes the sorts of things you can consciously remember and talk about, and they involve activity in the hippocampus and prefrontal cortex. The other sort of memory, nondeclarative memory, has to do with skills. Although many of Peter's declarative memories were blocked, he retained his nondeclarative memory of how to speak, walk, and use at least some abilities. This type of memory—and possibly the part of his brain that is involved in learning other people's abilities—involves activity in the amygdala and brain areas related to movement like the cerebellum and motor cortex.

Muscle Memory

Those areas are also important for Monica Dawson, who has a sort of "photographic" muscle memory. It would be interesting to see if the Haitian could block this part of Monica's memory—effectively blocking her use of her superpower.

"Like riding a bike, once you learn how, you never forget." Why is that? Unlike declarative memory—the sorts of things that you can say or describe with words, muscle memory has to do with skills. The Haitian's superpower doesn't affect those memories (although perhaps he could). That sort of memory involves parts of the brain that are very different from our declarative memories.

When we learn a skill, we may copy from a teacher and use some critical skills to hone our performance, but mostly, we just practice. Sheer repetition will allow us to memorize a procedure, whether it is how to ride a bike, how to kiss, how to write, or how to do the Funky Chicken.

This sort of muscle memory, or procedural memory, has to do with parts of the brain related to movement, like the motor cortex and cerebellum, as well as the amygdala. (Declarative memory, on the other hand, tends to use the medial temporal lobes.)

You know, I'm sure, that the Funky Chicken is a funny dance. But how do you know what is funny? You learn through trial and error, and more trial and error as your brain develops. Almost any response in a knock-knock joke seems hilarious to a preschooler—it is only after watching responses that a child learns to reduce (if not totally eliminate) absurd humor.

Trial and error. Repetition. Practice, practice, practice may be the way to Carnegie Hall, but this takes years of time. Who wouldn't want to shortcut that process? What if you could watch the Olympic trials and instantly become an Olympic-level athlete? Or play a little Guitar Hero and then give Jimi Hendrix a run for his money? We can wish it—but on *Heroes,* Monica Dawson can do it.

Real life, though, rewards real effort. In the classic musical *The Music Man,* the con man, "Professor" Harold Hill, claims that his students (whose families have prepaid for instruments not yet delivered) need only mentally practice to learn their parts on instruments they have never touched, much less played. It is a ridiculous proposition, an obviously ridiculous idea. And yet...many athletes visualize themselves going through motions, imagine their desired performance between practices. The difference is, the athletes are familiar with really performing their parts, and they still need to practice—and practice a lot—for the visualization to help. For Monica, however, seeing is as good as doing.

Monica doesn't have to practice to acquire gross motor skills, she just has to see someone else do it. This is wish fulfillment on the same level as Nathan's ability to fly. And that's a curious thought: if Monica can learn skills simply by watching other people exercise them, can she pick up the skills of other superpowered people? In other words, if Monica sees Nathan fly, will she also be capable of flying? Presumably, it would require her to have access to the same genetic code as Nathan, and to somehow activate the genes that enable Nathan to fly.

There's no reason to think that just seeing it would allow her to copy such an extraordinary skill. On the other hand, as someone who plays the instrument badly, I'd argue that playing violin well is also an extraordinary skill. Maybe as we see more of Monica, we'll see more of her capabilities.

Even though these characters—Matt, the Haitian, Monica—are fictional, they point out a lot of areas where scientists are pushing boundaries and exploring new frontiers in the effort to understand the mystery of what's going on in our heads.

Further reading

Norman Doidge, *The Brain That Changes Itself*, Viking, 2007.

William H. Calvin and George A. Ojemann, *Conversations with Neil's Brain: The Neural Nature of Thought and Language*, Addison-Wesley, 1994.

Chapter 11

EDEN, CANDICE, MAURY: PERSUASION AND ILLUSION; THE HAITIAN PART 2: BLOCKING POWERS

Every hero (or villain) is defined by her actions. But in order to become either, a person needs the freedom to choose her actions. Some of the scariest characters on *Heroes* take away that freedom: Eden McCain forces others to do what she says, like a Muggle version of the Harry Potter universe's Imperius Curse; and both Candice Wilmer and Maury Parkman (aka the Nightmare Man) create illusions to mislead the senses of people around them. That's pretty far out—real life isn't like that, is it?

In fact, real life *is* much like this—toned down a little bit, but maybe not as much as it would be comfortable to think.

Adamant Eden

Let's consider Eden, first. In the episode "Six Months Ago," policeman Matt Parkman stops a motorist, Eden McCain. In a normal encounter of this type, the policeman is the participant with power. He is in uniform and armed, he has been charged with the authority to question the driver, and he can apply a range of responses from letting the driver go (at best), to issuing a fine, to impounding the car and arresting the driver on the spot. Under certain circumstances, the police officer may even be entitled to physically restrain, injure, or kill the motorist. In other words, there was good reason for Matt to walk into the encounter suspicious of the driver and secure in the belief of his authority.

This traffic stop was not a normal encounter. Eden's first words to Matt tell him that she "kinda stole" the car and her attitude clearly doesn't reveal respect for his authority. She refuses an order from him to get out of the car. She suggests that he go back to his car and eat doughnuts. It takes two tries, but he does, in fact, stop giving her orders, go back to his car, and get doughnuts. As far as we can tell, he forgets about her as well.

What makes this scene so puzzling is not that people don't normally take orders or suggestions from another person. If that were true, then our lives would be full of much much more negotiation: bosses wouldn't be nearly so effective directing their employees, or teachers directing their students, or traffic cops directing cars. Negotiation is often time consuming and stressful. One reason that the military chain of command is so well defined and so strictly enforced is because negotiation could be fatal in time-sensitive crises. Whether the ends are good or evil, obedience is, at least, efficient. Every time we interact with another person, we navigate an entire web of permissible behaviors involving trust and obedience that are related to our self-concept, our idea of the status of the other person, our motivations, and the situation.

No, the confusing part about that scene in "Six Months Ago," is that we, the viewers, don't understand why Matt acted the way he did because we don't see the cues that would cause us to act that way. Maybe if he pulled over a car that contained the president of the United States and different words were exchanged, he might have ceded his authority, and acted in the same way we saw. Perhaps, if he pulled over a doughnut delivery driver who was rushing to the hospital, Matt might also have let the driver go, without ceding his authority—and that would explain why he developed a yen for doughnuts later. But as it is, Matt's behavior is inexplicable because it doesn't conform to our reading of the cues. Conclusion? Eden has some sort of superpower.

But you can convince someone to do a lot of things he wouldn't normally do if you provide the right cues. Stage magicians do it all the time. So do con men. So do advertisers. So do politicians. And cult leaders. And in Nazi Germany, millions of people who would normally have claimed that harming noncombatants and innocents was wrong nevertheless stood

STANLEY MILGRAM AND OBEDIENCE

In the early 1960s, a Yale psychologist named Stanley Milgram wondered if a person would be willing to inflict pain on someone else, without anger or hostility, simply because a third person told her to. He hired two actors and staged a play in which the subject was misled about the purpose of the experiment. (The ethics of experimental studies, which now require informed consent, have progressed considerably since then.)

In the basic experiment, two male volunteers come to a lab to take part in a psychology study about memory and learning. (The later experiments used subjects of both sexes.) Supposedly at random, one of them is designated a "teacher" and the other a "learner."

The stern-faced, white-coated experimenter explains that the study is concerned with the effects of punishment on learning. The teacher will read the learner a list of simple word pairs. When the learner hears the first word again, he must respond with the second word. The punishment for making an error is an electric shock. The shocks will increase with accumulating errors.

The learner is conducted into a room, seated in a kind of miniature electric chair, his arms are strapped to prevent excessive movement, and an electrode is attached to his wrist. After seeing this, the teacher is taken out of sight of the learner, into another area, and seated in front of an instrument panel with switches labeled with different voltages, ranging from 14 to 450 volts and signs with descriptions ranging from SLIGHT SHOCK to DANGER: SEVERE SHOCK. Each teacher is given a sample 45-volt shock from the generator before the experiment begins.

Now we come to the real experiment. Both the experimenter and the learner are actors. The electric chair isn't really wired to shock the learner. But the learner plays his part, beginning to protest as the voltages increase: first grunting, then complaining, then complaining vehemently and with increasing emotion, then screaming. The actor playing the learner controls a tape recorder on which a variety of groans and screams are recorded, so he isn't even subjected to abuse of his vocal cords.

The "teacher" doesn't know this. The real subject of the experiment is the person playing this role. The props and the actors and the scripts are designed to fool him into believing that he is actually inflicting increasing amounts of pain on the learner. The point of the experiment is to push the teacher until he balks at inflicting pain on the learner. When he does, the experimenter insists on continuing the experiment. (The experimenter has a script of responses as well. He insists that the shocks are not dangerous—despite the evidence apparent from the labels on the shock generator and the complaints of the learner.) For the teacher to stop the experiment, he must defy the experimenter's authority and refuse the order to continue.

When forced into such a situation, will the average person defy authority? There's no obvious drawback in doing so except the disappointment of the experimenter. Or will the average person become a willing tool of the white-coated experimenter, carrying out what appears to be a sadistic, even murderous, course of action?

Sixty-five percent of the people tested obeyed the demands that they continue until the end of the experiment. The percentage remained substantially the same (or higher) when the experiment was repeated in other surroundings and by other researchers. Milgram said, "Stark authority was pitted against the subjects' strongest moral imperatives against hurting others, and, with the subjects' ears ringing with the screams of the victims, authority won more often than not."

FURTHER READING

Stanley Milgram, "The Perils of Obedience," *Harper's,* December 1973, p. 62.

———, *Obedience to Authority: An Experimental View,* HarperCollins, 1974.

by and didn't protest the Holocaust—and many of them complied with the orders that made it possible.

The conclusion of Milgram's experiments? It seems that if "an expert" or "an authority" tells an average person to harm a third person, chances are that she will follow orders.

Milgram's experimental setup has a lot in common with the schemes of con men in the early twentieth century. If you've seen *The Sting* (and I recommend this movie), you see how the con artists set up a play—with a realistic stage, multiple actors, a director, and the unsuspecting dupe who can be depended on to play the part that those around him expect. The con artists convince the dupe to part with tens of thousands of dollars. Their base dishonesty is less upsetting to me, probably because I know I am watching a movie, than Milgram's experiments, which illuminate basic human tendencies that I would be more comfortable ignoring. Science doesn't give us the luxury of denial, however. We can argue about the meaning of Milgram's results (and people *have been* arguing about it for more than forty years), but the basic results have been duplicated, documented, and do tell us something about people in our society, at the very least.

The upshot? If Eden convinced us—by hypnosis or whatever method—that she wore the metaphorical equivalent of a white lab coat, chances are that she could command the average person to follow orders. Even orders that, like Milgram's, conflict with the person's beliefs and make the person deeply uncomfortable.

It wouldn't take much of a push. It might not even take a superpower—maybe all it would take is a lot of charisma. How many times have you met a person with great presence, or a magnetic personality, or fantastic ability as a speaker or performer? Someone who turns heads when she walks into a room, whose words seem to carry greater weight when she speaks, a natural leader, someone with "star power"? These people exhibit a combination of traits, including enthusiasm, grace, empathy, and self-confidence that draw people to them, and inspire obedience.

For example, Tom Brady, the quarterback for the New England Patriots football team, is remarkably charismatic. His first-rate performance on an excellent team (the Pats have gone to four Superbowls while Brady has been quarterback, and won three of them) helps, but it doesn't totally explain why he is so popular among sports fans. His charisma is undoubtedly aided by his high income and his tremendously attractive girlfriend, but his social intelligence also helps. Where his boss, Coach Belicheck, is renowned for both his civility (no bad-mouthing other teams) and grumpiness, Tom Brady shines. He is civil and gracious during interviews, powerful and effective on the playing field, as well as articulate, affable, and well dressed off the field.

Or take a counterexample. Here in Boston, ordinary folks treat politics as though it were a sport that is nearly as important as football. (It is seldom, however, treated as seriously as baseball.) We've had a mayor who is a good administrator—and not charismatic. Mayor Menino is fairly popular, he gets many notable things done, but he does not inspire his average constituent to public service. He's a good guy, but he's no Jack Kennedy. (In Boston, the Kennedy family and the Red Sox occupy pedestals of roughly equal height: it's hard to live peaceably with one's neighbors without admitting the preeminence of both.)

In a wider realm, it's election year in the United States as I type this, and the candidates for president have already been campaigning for many months. All the candidates must be at least vaguely credible, and be able to speak articulately about their positions, but a great deal of the campaign requires demonstrating charisma. Most people simply won't vote for someone they don't trust or like. (I'm not arguing that this is necessarily bad, but it doesn't have anything to do with rational thought.) At this level, charisma is a political necessity.

Although there are many disturbing trends in the current celebrity culture, many reasons why the most attractive person may not be the best person for every job, many reasons why a popularity contest is not the wisest way to appoint an administrator, for example, the desire to reward charismatic people is based on human nature. The allure of strong personalities, persuasion, peer pressure, and obedience are all programmed into our brains. And our brains are not entirely trustworthy.

A MIND OF ITS OWN

Our brains, our glorious brains, provide us with the ability to create dirty limericks, read body language, solve logic puzzles, memorize Shakespeare, watch TV, and go about all the other human activities, both splendid and mundane. Our brains are capable of fantastic self-awareness, which gives us the consciousness that separates man from the other animals, and is an apparently unique ability in nature.

Our brains are capable of abstract reasoning, and we derive many fruits therefrom. This isn't a particularly new idea. Shakespeare's Hamlet, surely the most depressed and indecisive prince ever to inhabit Denmark, expressed it this way: "What a piece of work is man! how noble in reason! how infinite in faculty!...in apprehension how like a god!"

We have good reason to celebrate our brains. But these same brains that we celebrate as the pinnacle of intelligent life? Those same brains that we depend on to reason and to discern truth? These same brains don't seem to be designed just for that, or maybe not even primarily for that. Our glorious brains are talented at deceiving us, at making us feel more important, more influential, more optimistic, and more talented than we really are. Our brains distort reality to protect our egos.

For example, the way a question is phrased has a great deal of control over how a person thinks about it. This can be far more subtle than the smear technique of obviously tainted questions like: "When did you stop beating your wife?"

In 1993, Eldar Shafir at Princeton University teased out the differences between the way we choose an option and the way we reject an option. The experiment posited a hypothetical child-custody case. Subjects were introduced to a set of facts about each parent. Then they were asked one of two questions, either "Which parent should have custody of the child?" or "Which parent should be denied custody of the child?" One parent was very average. The other had stronger "pros" but also stronger "cons" for parenting. If asked to award custody, more subjects chose the second parent. If asked to deny custody, more subjects *also chose the second parent.* Even given exactly the same information, the form of the question influenced the decision.

. . . continued from pg 241

This is only one of many ways in which we may think we're being logical, but are really being influenced by nonrational factors. So in the search for objective truth about our world, we're working with flawed equipment. The best we can do is be aware of the common ways in which people (and presumably us, too) delude themselves, and watch out for our most obvious biases.

In her excellent book, *A Mind of Its Own: How Your Brain Distorts and Deceives*, psychologist Cordelia Fine points out, "There is in fact a category of people who get unusually close to the truth about themselves and the world. Their self-perceptions are more balanced, they assign responsibility for success and failure more even-handedly, and their predictions for the future are more realistic. These people are living testimony to the dangers of self-knowledge. They are the clinically depressed."

FURTHER READING

Cordelia Fine, *A Mind of Its Own: How Your Brain Distorts and Deceives*, Norton, 2006.

Eldar Shafir, "Choosing versus Rejecting: Why Some Options Are Both Better and Worse Than Others," *Memory and Cognition* 21(1993), 546–556.

Milgram's subjects described themselves as good men, as kind people who disliked what they were asked to do, who were traumatized, despite or because of the fact that most of them complied with the experimenter's commands to inflict pain on an innocent person. Their brains may have had to do metaphorical somersaults to find a way to whitewash their actions, but to protect their egos, they managed.

This willingness to conform to the demands of authority—and in this case, it doesn't matter whether we place considered trust in the authority, or respond only from the feeling in our guts—has frightening implications for all of us. Will we answer the call of the next dictator or cult leader or con man who crosses our path? That can happen as easily as following a leader who inspires us to change our lives (and those around us) for the better.

Illusions

Having looked into this particular abyss of human nature, let's back away and look elsewhere. Both Candice and the Nightmare Man make people see things that aren't there. So do David Copperfield, Penn & Teller, and any number of stage magicians.

We can't necessarily believe our eyes. Or rather, the interpretations we put to the things we see are right most of the time, but consistently wrong in some cases. Vision is a complicated phenomenon. Our brains interpret what our eyes see using a series of assumptions that are usually correct but not at all rigorous.

What we see depends on what we expect to see. When we're around people, what we expect to see depends on social cues. Magicians manipulate social cues to misdirect their audience.

There are plenty of types of social cues. Body language is one: if someone points, you look where she is pointing, reflexively. Magicians can misuse the social cue to mislead, but usually the process of looking at where someone is

THE VANISHING BALL

In the animal kindom, gaze following is unusual. Humans are one of the few animals that pay attention to where other creatures are looking. (Dogs are another.) A group of scientists at the University College of London did an experiment based on gaze following, and found that, under certain conditions, people are quite willing to throw out the evidence of their own eyes.

The scientists had a juggler perform, one-on-one with the subjects. Her gaze follows a ball, and the subject's gaze naturally follows it as well. She fakes a throw, and looks at her hand. If that's all that happens, then the trick is fairly evident to the subject after momentary confusion. But if her gaze tracks where one would expect to see the thrown ball, then the results are different.

A majority of the subjects thought they saw the ball fly into the air, then disappear. This held true even if the subjects were informed beforehand about the trick. The combination of expectation and social cues overcame even the evidence of their own eyes.

There is some help for us, though. Watching the trick a second time, the subjects were able to ignore the misdirection. In this case, forewarned may not be sufficient to be forearmed, but at least it shows the value of direct experience.

FURTHER READING

Gustav Kuhn and Michael F. Land, "There's More to Magic Than Meets the Eye," *Current Biology* 16 (November 21, 2006), R950–R951.

pointing is useful; it communicates information. Similarly, if you see someone looking intently at something, you tend to follow her gaze. Have you ever seen someone on a crowded city street, looking up? The people around her look up as well.

But maybe there's a simpler explanation: maybe Eden, and Candice, and the Nightmare Man are all supertalented hypnotists. This could be a superpower. We know that hypnotism is real and that it induces changes in normal brain functions, but it's still a mysterious—if very useful—process. But it's not a free pass to mind control: one must, first of all, get the attention of the person or people to be hypnotized, and second, some people are more easily hypnotized than others.

Blocking powers

But what about the Haitian? In addition to his "anti-mind-reading" ability of removing memories, he also has the ability to block superpowers when he's near a person and focusing on them. His superpower even worked on Peter, who was immune or shared in nearly everyone else's powers.

Maybe this isn't so strange. If the Haitian can wield his power with sufficient delicacy to remove some memories but not others, then he's clearly talented at negotiating with the mind or brain. Why not suppose he could do the same trick on the part of the brain (or the mind) that is active when superpowers are used?

(What is more curious is that we don't know how the Company developed drugs that block powers. If we know what the drugs do, wouldn't that provide a great clue about where they are acting and what unusual activity occurs where in the brain when a superpower is used? Maybe that's part of the testing that the Company has been doing, as well as inserting an isotope tracer, on abducted individuals with superpowers.)

True, when Hiro lost his powers briefly, the implication was that this was a psychological block caused by grief or lack of confidence rather than a physical problem. But the Haitian doesn't stop our characters from believing in themselves—he just stops them from using the superpowered talents that they've developed.

There is a technology that can knock out—as well as stimulate—certain brain functions. It doesn't require surgery or drugs. Transcranial magnetic stimulation (TMS) can make muscles twitch, disrupt the ability to talk, and (perhaps) treat depression. Perhaps it could also zap the part or parts of the brain that control superpowers?

TMS depends on the ability of a changing magnetic field to create eddy currents. A magnet creates a magnetic field around it. If you move the magnet (or otherwise change the field), electrons in nearby conductors will move, setting up eddy currents. That's physics, and it's been well understood as Faraday's law of induction. Now, the neurons in our brains contain axons, which are electrical conductors. So if you create a strong, changing magnetic field near a person's brain, it will cause some of the neurons to fire without cause, or stop others from firing when they normally would.

TMS works by putting an electromagnet next to the person's scalp and generating short magnetic pulses. The pulses pass through the skull (hence the *transcranial* in TMS) and stimulate parts of the brain. The field strength isn't particularly powerful—about the same as an MRI scanner's magnetic field.

Interfering with the entire brain wouldn't be all that interesting, but by focusing the pulse, we can more or less locate what neurons are attached to what muscles. Researchers have used TMS to locate where neurons reside that make abdominal muscles contract. These neurons were in slightly different places in the brains of different men, but the researchers were nonetheless able to make all of the subjects' six-packs twitch.

Back in 1994, researchers in London found that by stimulating a certain area of the brain, they could interfere with the subjects' ability to move their hands. The effect was very brief—but a few milliseconds longer than the duration of the pulses.

Muscle control is an obvious way of checking what a magnetic pulse can do, but what about less physical responses? By combining the TMS technique with a brain-imaging method (like those mentioned in the chapter on Matt), we can affect one area of the brain and see what changes in other parts of the brain. A number of researchers are using functional magnetic resonance imaging with TMS for just that purpose.

If the pulses are repeated at fairly high frequencies of, say, more than five per second, the pulses seem not to inhibit brain activity but to stimulate it. For clinically depressed people with particular parts of their brain that are less active than normal, TMS might offer a kick-start. In some experiments, TMS has improved the mood of the subject for days or weeks after stimulation. But researchers are still trying to figure out what parts of the brain to stimulate, what dose works best, why it works at all, the best frequency and ways of avoiding side effects, and whether there are long-term health effects from the process.

Maybe the Haitian's ability to suppress other people's superpowers is based on being able to somehow project magnetic fields into parts of their brains? If that is the case, then his power is more like Micah's than Eden's: based more on electromagnetism than on personality.

However he does it, the Haitian's powerful abilities benefit whomever he works with. His motivations are obscure, but at least near the end of Season Two, his heart seems to be in the right place. The difference between being a villain or a hero depends not on what you can do, but on what you choose to do, and on how you choose to do it, as we will see in the next chapter when we look at the two characters with the most similar set of powers and the very different ways they use them: Peter Petrelli and Sylar.

Further reading

Steven Pinker, *How the Mind Works*, Norton, 1997.

Lauren Slater, *Opening Skinner's Box: Great Psychological Experiments of the Twentieth Century*, Norton, 2004. A very personal take on the subject, but also very readable.

Alison Motluk, "Mind Readers of the 21st Century," *New Scientist*, November 12, 2005.

David Maurer, *The Big Con: The Story of the Confidence Man*, Bobbs Merrill, 1940. How confidence men used psychology to steal many thousands of dollars from their victims. Also, a really fun read by a linguist. Reprinted in 1999 by Anchor Books.

Chapter 12

PETER AND SYLAR: WHAT MAKES A HERO?

Are you a hero? Are you a villain? Are you a victim of cir-
cumstance? Are you "just a cheerleader"? We are each the
heroes—or at least the protagonists—of our own lives. Our
actions, even more than our intentions, determine whether
we are counted among the great and good or among the
scourges of our kind.

And that is the basic difference between Peter Petrelli
and Sylar. Their powers are very similar: both are capable of
absorbing other people's superpowers. If, as we've posited,
the powers are genetically based, both of them must be capable of rearranging their genes to express new arrangements.
Like the cuttlefish changes its shape and skin, Peter and Sylar

appear capable of genetic gymnastics that are a superpower in themselves.

The effect is similar, but the method of determining how to change is different. Peter is a woobie: he cares about people, he is deeply connected to his family, he devoted years of his life to nursing people. He is loving and lovable. Sylar, on the other hand, is isolated and selfish and sociopathic. Syler is about as cuddly as a cockroach, and apparently as durable as one.

Peter's power

Peter's best intentions didn't stop him from almost causing a huge explosion in New York City. Nevertheless, his intentions are uniformly good, which at least makes him sympathetic. His method of replicating powers seems to require only proximity, and then expressing them requires feeling empathy. (Is his talent triggered by part of the brain that handles emotion?)

What do we mean by empathy? For a word describing emotion, empathy is a fairly recent concept, having been coined only in the 1850s. It means, roughly, the ability to recognize someone else's emotions and feel them oneself. Feeling empathy for another person doesn't necessarily evoke sympathy or the desire to help her.

In our brains, empathy may be related to mirror neurons.

For example, we demonstrate empathy if we recognize that someone we see feels anxious, and there is more activity in the pathways in our brain that fire when we are anxious. Our emotions copy the other person's emotions.

In Philip K. Dick's novel *Do Androids Dream of Electric Sheep?*, empathy defines who is human. A test for involuntary empathic reactions is used to distinguish who is human (as opposed to an android). Yet animals also appear to empathize with their fellows.

Perhaps Peter's brain not only mirrors the emotional state of the people he empathizes with, but somehow also mirrors their mental activity when they use their powers. How he could possibly perceive that they are using powers (which he has never used) and mimic them, is another question entirely.

Does the nose know?

We can't tell who on *Heroes* has extraordinary powers based on merely looking at them. Unlike comic-book superheroes, our characters don't wear spandex or have wings sprouting from their back or other obvious deviations from the anatomical or sartorial norm. Their voices usually sound normal. But maybe they smell subtly different?

In other chapters, I discussed the possibility that electric or magnetic fields might transmit information—allowing

Matt to read minds, for example—but there's another way that people can transmit information when they are close to one another. It is based on the fact that people smell.

We all release gases through our skin (not to mention from either end of our gastrointestinal track) continuously. It is involuntary. These gases provide information about our metabolism, our diet, and our health. They also provide information about our sexual state: both maturity and interest.

Scent and mood seem intricately entwined. Compared to many other animals, our human sense of smell is dull. But even so, our noses are remarkable detectors. We can distinguish more than ten thousand different smells. Actually, what we are doing when we smell an odor is sensing a chemical in the air, often at remarkably low concentrations. For some especially pungent chemicals such as, say, those emitted by an angry skunk, we can detect a few parts per billion. In other words, suppose a skunk has sprayed in an area we are in (hopefully in the open air). The odoriferous compounds, called thiols, have dissipated somewhat. But if they make up only one-billionth of the volume of the air we breathe in, we'll still be able to smell the skunk.

What are our bodies doing when we smell either the angry skunk or the sweet skunkless air? We have special neurons that line the insides of our noses, each of which detects certain chemicals. (Except that's still misleading: it's not a case of having one receptor, or one neuron, designed to smell skunk, while another identifies roses. Recognizing a smell depends on the pattern across all the neurons in the olfactory bulb.) Each of the hundreds of receptors are encoded by a specific gene, which means that a fair amount of the genes in human DNA—something like 2 percent—is concerned with whether something stinks like a rotten egg or is scented like a rose or smells like our favorite person. Well, sort of. Many of the genes that seem to encode odor receptors...apparently don't. It looks like only about 350 genes actually code for functional receptors. Maybe that explains why other animals, from bloodhounds to sharks, seem to have much sharper senses of smell than humans.

Although this fact doesn't have a prominent place in our current society, we each have a unique and recognizable smell. Not only that, but we can smell emotions. The old saying that dogs can smell fear is literally true: both dogs and horses are very sensitive to the smell of fear in humans. And humans can smell the difference between armpit swabs taken from someone who is happy and someone who is sad. (Women were more effective than men at distinguishing between the two, and other studies show that women, particularly women of reproductive age, have a more acute sense of smell than men, and that the smell sensitivity of most women varies across the menstrual cycle.) In addition to fear, happiness (or at least contentment) and sadness, sexuality, and memory are intricately linked to smell.

Pheromones

And then there are other airborne chemicals that may affect us without having a noticeable odor. Pheromones are airborne chemical signals that are released by one individual of a species; when other members of the same species detect the signal, they can change their behavior or how their bodies are working (their physiology). Most of the research on pheremones has been done on insects, which are, relatively, simple. Insects can signal danger, or sexual interest, or the availability of food. More subtly, a pheromone given off by a queen bee will make other females sterile. Many pheromones have been isolated and identified for different types of insects, and some mammals, including pigs, mice, and hamsters. A mysterious structure in the nose, called the vomeronasal organs, appears to detect pheromones, and it's different from the rest of the olfactory organs. The signals from this organ travel to a second olfactory bulb in our brains (rather than our main one). In rodents, the signals travel to parts of the brain that control reproduction and maternal behavior. If humans work the same way, then the signal sent when we sniff a pheromone never gets close to the cerebral cortex, and thus may never get to the conscious part of our brains.

That's the theory, at least. Thus far, no pheromones have been identified in humans. Despite this, some companies advertise scents containing pheromones, usually with the suggestion that they will make you irresistible to members of

MARTHA McCLINTOCK AND MENSTRUAL MANIPULATION

As an undergraduate at Wellesley College, a young woman named Martha McClintock noticed something that had, perhaps, been known anecdotally but hadn't been reported in scientific literature: women who live together tend to sync up their menstrual cycles. After doing more research (on her dorm mates, among others), she theorized that pheromones were responsible for the synchronization. Her research was published in 1971 in the (very prestigious) journal *Nature*.

But it wasn't until 1998 that she (now a professor of psychology at the University of Chicago) and Kathleen Stern published the results of an experiment—again, in *Nature*—that showed evidence of pheromones in humans. Specifically, they discovered that women's underarm sweat could influence the hormonal response of other women exposed to it—even when the compounds were odorless.

They again investigated menstrual synchronization. The researchers swabbed the underarms of women at two distinct points in their menstrual cycle. Each swabbing pad was cut into four pieces and frozen. The pieces of the odorless pad were rubbed under the nose of the women test subjects. In most of the recipients, the release of hormones that trigger the steps of the menstrual cycle in the recipients was either accelerated or delayed to more closely match the state of the donor test subjects.

So people—or at least the women in the study—are capable of responding to a chemical signal in some way that isn't dependent on smells they can identify. Even if we don't know what the chemical is, we can see it acting on our bodies in ways outside of our conscious control.

Since then, McClintock and others have continued experimenting on human response to other people's smells, and have found other examples of chemical signaling on hormones related to sex and child rearing.

FURTHER READING

Kathleen Stern and Martha K. McClintock, "Regulation of Ovulation by Human Pheromones," *Nature* 392 (March 12, 1998), 177–179.

the opposite sex. Most of them don't specify what species the pheromones are from.

But even if we haven't isolated chemicals that we can identify as pheromones, we do seem able to sense chemicals that transmit information, without consciously being able to smell odors.

What if Peter has a super nose for superpowers, picking up information about the powers of people around him—enough to allow him to mimic them—based on smell alone? Because smell receptors are based on genes, you can't smell certain things (camphor is one example) if you lack certain genes. Peter's superpower may include having extra genes—or activating normally ignored genes—to allow him to smell things that nobody else smells. Peter might not consciously realize that he's been "smelling powers," but perhaps, in some way, the smell of a power provides him with enough information to change his DNA to enable the same power.

There is one bug in this theory: what about the Haitian? If Peter could absorb the Haitian's power, he ought to be immune to it—the same way Peter could see Claude when Claude was invisible to everyone else. But the Haitian manages to block Peter's powers and then takes his memory away, at least temporarily. Perhaps the Haitian, in addition to his other powers, doesn't emit pheromones or whatever signal Peter detects?

And Peter detects a lot. He picked up a lot of powers including the ability to travel in space and time and to see the future, and the ability to create radioactivity; healing he got from Claire, invisibility from Claude, telepathy from

Matt, shooting lightning from Elle and flying from Nathan. And then, when he encountered Sylar in the episode "Unexpected," he was exposed to a whole raft of powers all at once, most notably telekinesis. If Peter is sniffing them out, Sylar's collection of powers must have smelled like a Yankee Candle Factory store.

Intuiting systems

Yes, and what about Sylar? Whereas Peter identifies with other people and absorbs powers almost as a side effect of his empathy, Sylar appears to be motivated by the desire for power. Originally, his power was the ability to analyze watches (and presumably other machines) to understand both how they worked and how they failed to work. As Gabriel Gray, he seemed to know, intuitively, how mechanical equipment works. When his ability extended beyond just these machines to understanding the workings of the superpowers of the other "evolved humans," his thirst for supremacy led him to kill the first other evolved human he encountered, and to steal his superpower of telekinesis in the episode "Six Months Ago." Now his modus operandi, followed throughout his rampage, is to find a person with a superpower, remove her brain, and gain her power. In Sylar's mind, killing people appears to be an unimportant side effect of taking their brains.

261

Really, it's quite a jump for Sylar to progress from understanding mechanical systems to understanding something mysterious about superpowers that is physically expressed in the human brain. He doesn't have an affinity for complex mechanical systems or he would be an evil mastermind gadgeteer. Sylar also doesn't show any sign of being able to understand electrical and electronic systems, like Micah does. He doesn't appear to comprehend chemistry. He shows some signs of understanding psychology as he manipulates Maya and other victims and adversaries, but he's not brilliant at it. Maybe he doesn't have to be?

If Gabriel had chosen to use his original superpower for the common good, he could have become a brilliant mechanical engineer, or a troubleshooter in any one of many industries, or a fantastic urban planner. Instead he turned to evil.

We don't know exactly what Sylar does with the brains or why he needs them to steal power. We know he is uninterested in the brains of nonevolved humans since he doesn't seem interested in killing them and he abandons Jackie Wilcox's brain in the episode "Homecoming." (Perhaps he could have looked at Jackie's brain to discover cheerleading abilities?) We have seen that the brain needs to be in good working order (dare I say, "fresh"?) for it to be useful to him—we know that when Eden McCain shot herself in the head, he was unable to obtain her power.

Perhaps by merely looking at a brain, Sylar can deduce and replicate the superpower of the person. That would be a power that many neuroscientists would love to have.

THE BRAINBOW

One of the problems in brain research is that the neurons that connect to one another all look the same. Imagine trying to sort out a tangle of several different balls of yarn: it's much easier if the balls are of different colors and textures, much harder if they are all the same. In the case of our neurons, each one has gray parts and white (myelin-covered) parts, and they are very very tangled. In order to see them clearly, in the past, they have had to be stained—which is usually done to thin sections extracted and stabilized on a microscope slide. Not so helpful for looking at brains in action.

As a first step toward untangling the mess, researchers at Harvard University figured out how to color-code mice brains. Harvard professors Jean Livet, Joshua R. Sanes, and Jeff W. Lichtman added some genes to lab mice that cause their brains to make fluorescent proteins. But not just one color—by activating multiple fluorescent proteins in neurons, they can create up to ninety colors!

The new method labels each nerve cell with its own mixture of blue, yellow, orange, and red fluorescent proteins, more or less at random. Then, by using confocal microscopy, the patterns of neurons show up, vividly painted.

The researchers used this method to look at mice brains for nearly two months, and were able to watch the neurons reorganize over time. The actual results from using glowing-brain mice are just beginning to come in.

FURTHER READING

Jean Livet et al, "Transgenic Strategies for Combinatorial Expression of Fluorescent Proteins in the Nervous System," *Nature* 450 (November 1, 2007).
Roger Highfield on Telegraph.co.uk (the Web site of the British newspaper *Daily Telegraph*) explains the work in more human-readable prose than the *Nature* paper: "Brainbows Offer Unique Colour Brain Map," October 31, 2007, at http://www.telegraph.co.uk/earth/main.jhtml?xml=/earth/2007/10/31/scibrain131.xml.

Is Sylar what he eats?

Another, more alarming, possibility is that Sylar takes a line out of zombie movies and actually eats part of his victims' brains. Apart from breaking the long taboo against cannibalism, what might Sylar gain from eating nerve tissue?

Humans are omnivores—we eat all parts of plants, and just about any part of an animal short of hair and hooves. So, in an unsentimental worldview, it's no wonder that, occasionally, people eat one another. Sometimes the reason for cannibalism is survival, à la the Donner party. Monty Python's "Lifeboat Sketch" makes light of this—with the castaway sailors arguing about whom they should eat—but in some ways human flesh is similar to animal flesh and it does provide nutrition. The movie *Soylent Green* made this pretty clear, as did *The Twilight Zone* episode "To Serve Man."

Some species of animals are cannibalistic as a matter of course, like the praying mantis. And even among those that would not normally eat their own species, exceptions occur. For example, livestock in some countries are fed with remains from other members of their species.

But there are also very good reasons for not eating one's peers. There are all sorts of disease-causing bacteria and viruses that can transfer as easily within individuals in a species as between species. And larger organisms, like tapeworms and other relatively complex parasites, can move to a new host that eats the previous host.

Obviously, the possibility of transmitting diseases from the dead to the living animal is a concern, so both heating and sterilizing remains is standard practice for almost any meat that we eat. For livestock, remains are also usually sterilized. (Humans eating steak tartare and sushi are exceptions.)

At other times, cannibalism is a cultural event. Although cannibalism, also called anthropophagy, is usually taboo, in a few societies it has been acceptable. Among the Fore people of New Guinea, ritual cannibalism was, at least for a time, a respectful funerary observance. And although the Fore did cook the dead before eating them, they still got a mysterious disease called kuru.

Kuru is a horrible and bizarre disease. Victims suffered progressive neurological degeneration: first muscle weakness, then dementia, then death. The disease is always fatal. There was no obvious route of infection. The first researchers thought that it might be inherited, until that was disproved. It was then theorized that kuru was a virus with a very long incubation period. But no antibodies could be identified, which one would expect if the pathogen was either a virus or bacteria. Brain tissue, inspected after death, had a number of tiny holes in it, like a sponge.

Other diseases show similar signs, and are called spongiform encephalopathies (SEs). The most infamous may be bovine spongiform encephalopathy (BSE), better known as mad cow disease. Roughly 160 people in Britain have been diagnosed with a similar disease, called variant Creutzfeldt-Jakob disease. Both diseases also resemble a disease of sheep and goats, called scrapie.

These SEs drove infectious disease experts crazy because they couldn't figure out how they transferred from one person or animal to another. For one thing, the incubation period seems to be no shorter than two years, and we're still trying to figure out how long it can last—recent estimates suggest that the time between infection and showing symptoms may be as long as fifty years!

A second problem is that standard methods of killing pathogens (disease-causing organisms) didn't work on the BSE samples. Infected nerve tissue was treated in a variety of ways before being fed to animals in laboratory tests, and the animals kept dying. (Actually, infected tissue from sheep, in the form of scrapie, was used for these tests. Mice, it turns out, can get scrapie.) Cooking didn't destroy the pathogen. Shining UV light at it, or soaking it in an antiviral fluid—both methods designed to disrupt the nucleic acid sequences in the sample—didn't stop the tissue from infecting the test animals.

That scared the heck out of the researchers. It scares the heck out of me! And it provided a strong impetus for people in Britain to stop eating any meat at all. This was a reasonable response, although given the potentially very long incubation period, it might be a case of closing the barn door after the horses are already out.

There seemed to be only two possible explanations: either

the pathogen has some incredible protection for its nucleic acid sequences, a sort of "superbug" that isn't affected by any of our standard safety procedures…or it wasn't based on nucleic acid sequences.

But hold on a minute! The accepted wisdom was that any infectious agent included nucleic acid sequences (DNA or at least RNA) in order to replicate. Certainly, DNA and RNA are simple and robust and very widely used ways to replicate organisms.

But what if the disease isn't caused by an organism? (On the other hand, it obviously spreads, which requires replication, and what is an organism except something that can self-replicate? This is a sort of chicken-and-egg dilemma.) The disease does seem to spread—although it may also spontaneously appear. If it spreads, maybe it is caused by something even simpler than RNA, some replicating mechanism that can withstand the normal sterilization methods, something without nucleic acid in it.

We now think that this is what's going on, this single method explains how kuru and BSE and related diseases spread. Although there are plenty of questions still left open, as near as we can tell, all the spongiform encephalopathies are caused by a simple protein. And it's not even an unusual protein—it's one that exists in the brains and nerves of many animals.

The culprit protein has all the same atoms as the normal form, but it is folded funny. Just that single change, that difference in how the molecule bends, is enough to kill the host if it propagates.

It's called a prion.

Prions have the remarkable ability to cause a normal protein to change shape, and turn into a misfolded protein. Then the previously normal protein can go off and cause other proteins to change shape. Tada! Self-replication. Self-replication without nucleic acid. Also: infection.

It doesn't take a lot of proteins to start the avalanche of infection. In tests on cattle, less than a teaspoon of infected nerve tissue from a cow with BSE was enough to cause BSE in another cow.

BSE can occur even in animals as simple as yeasts. More complex animals, like us, depend on our digestive tracks to defend against many potential infections. Although digestion breaks down a lot of problem organisms, proteins (correctly or incorrectly folded) can get through, transferring from our guts into our blood.

Part of the problem in first understanding and then in battling this disease is that we still don't know a lot about proteins. Although we've identified a lot of important proteins that go into running our bodies, we're still figuring out what shapes they take normally. And we don't know how normal proteins usually find the single, correct, folded shape that they take.

We don't even know why prion diseases are so rare. They

can occur spontaneously. The tendency toward prion diseases can be inherited. Or they can occur via infection, such as from one caused by eating infected animal tissue. What stops them? We don't know.

The possibility of contracting a prion disease probably doesn't bother Sylar, though. Certainly the FBI is a more immediate risk to him than a slow-acting disease.

And in the end

No matter how they absorb the unique powers of people around them, both Peter and Sylar appear to mimic or gain their powers by making changes to their own DNA somehow. And that comes back to the basic question of what the superpowers are, what we mean when we talk about activating genes, and evolution. Is it merely a case of using potential already held within their—and our—DNA? Or do they transcribe foreign genes into their own DNA, becoming self-transgenic animals, something brand-new under the sun?

Part of the delight of *Heroes* is that the characters might be just like us. You may not have a superpower. Then again— you might! Either way, the ethics of one's choices matter. The power of your mind, your intentions, and especially your actions can make you a hero. The decisions that you make can be made selfishly or for the greater good. The work you do—as a scientist, or a policeman, a teacher, or a politician—can be done well or badly. What you do with your voice and your vote and your dollars makes a difference, both in the story of your own life, and in your community, and in the world.

Chapter 13

THIS IS HOW
HEROES WORKS

Ordinary people encounter the extraordinary. They question it, poke at it, talk about it, make theories to explain it, test it, and learn how to use it. It is what we do.

This is what we do: a child discovers that a prism breaks white light into colors, and tilts it to see how the colors move. A scientist discovers that her experiment is giving off way more heat than she knows how to explain, and she changes variables to see what happens next. This is also what our superpowered characters do: Claire discovers that she can heal ridiculously quickly and tests it by jumping from heights, Hiro teleports and teleports again for practice, Matt hears thoughts and strains to understand what he's hearing, Peter struggles to control a panoply of unexpected abilities.

This is what we do: we each have priorities, the people and things and causes that we care about. Noah Bennet fights to protect his family. Niki and D.L. struggle to maintain their marriage and take care of Micah. Claude spends decades hiding, but risks his cover to teach Peter (in Claude's distinctively abrasive way) how to control his powers. Nathan runs for Congress. Ando follows Hiro to America out of friendship. They stumble down destructive paths sometimes, they make mistakes. Their goals shift sometimes, like the way Matt loses his marriage but finds himself parenting Molly.

As Hiro Nakamura tells us, "You don't have to have superpowers to be a hero."

On the other hand, you don't need superpowers to be a villain. You just need selfishness. Sylar and the Company pursue power without restraint. Sylar's one priority is gathering abilities to himself, murdering without hesitation. The Company, whether it was originally intended for noble purposes or not, was driven by the fear of the founders to become a malevolent organization that kidnaps and imprisons and even kills people with superpowers.

Our heroes—superpowered or not—take responsibility for themselves and for others. They oppose the selfish Sylar and the destructive Company with the tools at hand and the talents they've been gifted with.

There's one character I haven't discussed yet, one charac-

ter who is near and dear to my heart: Mohinder, the scientist. Mohinder cares deeply about the people around him, and he also cares deeply about the science pioneered by his father. To him, science provides an answer, an explanation about the universe, and about how ordinary people have suddenly (or perhaps not so suddenly) begun to exhibit extraordinary abilities.

And that's what science is all about: it is an explanation for what we see around us, an explanation that can be shared, and that—to the best of our ability right now—we can't prove false.

The scientific method sounds cold and analytical, but each scientist is passionate about her work—she has to be, to put up with the obstacles that inevitably get in the way. On *Heroes*, the lines between right and wrong get confusing sometimes. The real world is like that, too. The real world of science experiments is messy, obscure, confusing. Equipment doesn't work or is imprecise or (worse) inaccurate, equations are refractory or underspecific or overspecific, under precisely controlled conditions the organisms do whatever they please, experiments may have unexpected errors in design or in the way the results are interpreted. Or, our entire intellectual model describing some aspect of the universe may rest on an incorrect assumption: that was the case before we learned that the planets (including our own) revolve around the sun, as well as before quantum mechanics taught us that the universe is statistical but not deterministic.

Even Einstein, as smart and insightful as he was, had difficulty accepting that—on a very small scale, at least—the energy and location of tiny fragments of matter are not only unknown, but unknowable. And yet, well over half a century of experimentation has shown us that it is apparently so, and we make allowances for and even use quantum uncertainty in today's technology.

Progress in science is not certain. And that's what makes it so exciting. In a delightful book called *Underwater to Get Out of the Rain,* marine biologist Trevor Norton writes, "Scientists find out by *doing,* not by being told. Existing ideas and facts are merely signposts to the future and sometimes misleading ones at that. To linger too long on what is already known, however interesting that may be, is to be distracted from the business of science, which is not just the accumulation of facts, but the pursuit of the new and, if you are lucky, the unexpected."

Mohinder is grappling with the unexpected. A man who can fly is unexpected. But if it's possible to fly, then it's possible to describe how he flies. The same is true of a girl who can locate people from a huge distance, or a boy who can talk to (and receive answers from) traffic lights.

Hiro's quest is to save the world with a sword and a brave heart. Mohinder's tools are less dramatic—a lab and DNA-sequencing instruments and a vaccine—but his goal is just as noble. If the explanation for the abilities of the characters lies in their genes, Mohinder's quest is to unravel the spools of DNA, develop an explanation, test it, *and then share that knowledge.*

We need both Hiro and Mohinder. We need inspiring stories and the building blocks of science.

It's difficult to stop the process of investigating the extraordinary, of poking at mysteries. We are curious creatures, living in a universe full of wonders, beauty, fascinating phenomena, and amazing complexity. This is what we do. The process of science is a powerful tool for learning to appreciate how it works.

In order to meet the challenges of the present and future, we need to understand the world and the problems before we can find solutions. Respecting the truth and wonder of the physical world by practicing science is, in and of itself, good for mankind.

And also, it can be a whole lot of fun.

Further reading

Trevor Norton, *Underwater to Get Out of the Rain: A Love Affair with the Sea,* Da Capo Press, 2006.

Alan Lightman, *A Sense of the Mysterious: Science and the Human Spirit,* Pantheon Books, 2005.

(132) DNA damage radiation